Detlef D. Spalt

A Brief History of Analysis

With Emphasis on Philosophy, Concepts, and Numbers, Including Weierstraß' Real Numbers

 Birkhäuser

Detlef D. Spalt
Darmstadt, Hessen, Germany

ISBN 978-3-031-00652-4   ISBN 978-3-031-00650-0   (eBook)
https://doi.org/10.1007/978-3-031-00650-0

Mathematics Subject Classification: 01Axx, 97-xx

This book is published under the imprint Birkhäuser, www.birkhauser-science.com by the registered
company Springer Nature Switzerland AG
The registered company address is: Gewerbestrasse 11, 6330 Cham, Switzerland

to Juliane
in memory of 19 May 1990
and 27 January 2017

Preface

For Whom Is This Book Written?

This book will be of interest for all those who ask themselves: what is mathematics? how is it accomplished? from where did *its concepts* originate?

Some *proofs* are shown—not the celebrated ones but the fundamental, the important ones. Among them is one which does not add up but ends in controversy (Chap. 5). One of the renowned and favourite proofs of devotional mathematical literature is called into question and proven to be *only partially conclusive* (Chap. 14). Not uplift but knowledge is our aim, not well-being but thought.

Which thought led up to today's mathematics (especially: modern Analysis)?

Who Can Understand This Book?

The understanding of this book does not require an academic degree in mathematics. However, you need some acquaintance with the central concepts of Analysis (function and series, continuity and convergence, differential and integral). The details are always given on the page.

What Is at Stake?

This book shows in a historical way:

- How and by what means was Analysis accomplished? (the *formation* of mathematics, the philosophical aspect)
- How do the objects (the concepts) of Analysis develop and change? (the *modelling* of mathematics, the conceptual aspect)

Two new *mathematical* discoveries are added:

- Today's common Analysis may easily be changed and given another shape—
  which is to say: some other theorems become valid. Our theory makes up the
  archetype of modern Analysis, which was created by Cauchy.
- Besides the two constructions of the real numbers known today, there exists
  a third version, which is completely different. It was invented by Weierstraß
  and taught in his foundational analysis course but was not understood and,
  consequently, not passed on.

This are two *historical* issues as well as two items of *mathematical* news: the history
of mathematics is able to present novelties even to mathematicians!

Who Has Contributed?

My thanks go primarily to a dozen master's students of mathematics who stimulated
me by their interest, their vigilance and their criticisms (Mr. Marco Pavić was
especially engaged) in the winter semester 2016/17, as this book is based on the
lecture notes written after each session, and subsequently also to Frankfurt am Main
university for granting a fifth—unpaid—teaching assignment in the Department of
Didactics of Mathematics. These students demonstrated that there still exists an
ongoing student interest in the history of mathematics, even though there is no
proper room for this specialization in the mathematics curriculum after the Bologna
Process.

Inevitably, my special thanks go to Prof Dr Jürgen Wolfart for his organizational
contribution as well as for his thoughtful hint to look at the recently discovered
student's lecture notes from Weierstraß' lessons in winter 1880/81. Without this
fact, these notes would never have come to my attention.

History and didactics of mathematics have in common that they both focus and
reflect on present mathematics. This may cause a mutually fruitful relationship
as has happened in this case. In particular, I am indebted to Prof Dr Reinhard
Oldenburg, now Augsburg, and Prof Dr Harald Riede, formerly Koblenz, for
their reference to the formula concerning the concept of "continuous derivability"
(p. 144).

As always, I am indebted to my very persistent friends Hassan Givsan and Bernd
Arnold for their everlasting willingness to criticize my tentative formulations in
respect to essence and linguistic quality, thereby helping to clarify and to improve
the book. Finally, Juliane engaged in a final criticism with élan and success: thank
you!

However, without the curiosity of Mrs Iris Ruhmann from Springer, her open-
mindedness and her courage to include this book in her programme, these lecture
notes, as usual, would have been lost as a single document in the university library,

thus remaining unknown to the public. Mrs Ruhmann is entitled to the merit for this publication, and I thank her very much for that.

Darmstadt, Germany   Detlef D. Spalt
15 June 2018

Preface to the Translation

This translation includes one additional mathematical and one further historical novelty:

- The *mathematical novelty* is Weierstraß' concept of real number (in Chap. 12). Although I had embarked on the study of the newly found lecture notes taken in Weierstraß' lessons of winter 1880/81 in preparing my lecture in winter 2016/17, I failed to understand it correctly. It was only in spring 2017 when I succeeded in unravelling the puzzle and understanding Weierstraß' fundamentally new idea of a number concept. It is of such a revolutionary essence that this book seems to be the first documentation of this mathematical idea in print in English—as no referee of an international first rank journal was able to understand and accept it for publication within 5 years.[1]
- The *historical* novelty is the unearthing of Traugott Müller's schoolbook from 1838, wherein Müller clearly defines the (positive) irrational numbers in the very same manner we usually call "Dedekind-cuts" (in Chap. 13).

As to the burden of translation, I am indebted to the longstanding pair of friends, who taught British A-level Mathematics, Juliane Horn and John D. Smith (London), who transformed my often colloquial German into decent English. Clearly, no one could have done better than they did!

Darmstadt, Germany   Detlef D. Spalt
7 December 2021

---

[1] See Detlef D. Spalt 2021, The concept of real number, following Karl Weierstraß. URL https://www.detlef-spalt.de.

Detlef D. Spalt 2022, *Die Grundlegung der Analysis durch Karl Weierstraß – eine bislang unbekannte Konstruktion der natürlichen und der reellen Zahlen.* (Springer).

Contents

Introduction: The Four Big Topics
of This Book

All those who are not yet familiar with the history of mathematics are advised to start directly with Chap. 1. However, those who have already had a first encounter with the history or with the philosophy of mathematics, will get a first orientation by this detailed overview of the four topics we discussed on p. vii. From p. xviii to p. xx the shortest possible history of analysis is given.

The *Configuration* of Mathematics—or: Designing Mathematical Theories

From the works of the classical Greek philosophers we know: mathematics is conceptual thought. Whilst Babylonian accountants put up with calculations for millennia (thereby producing the abstract *object* "number"), the Greek philosophers invented the critical discourse, *i.e. concepts*. With Euclid (*c*-300), mathematics took the form it still has today: *definition–theorem–proof.*

Without definition and without proof, there is no theorem. This book tries to shed light on those two areas of mathematics, which are usually overlooked: the *definitions* (or: *"concepts"*) and the *proofs* (which rely on those definitions). It turns out that these are the decisive issues.

To Define Is Hard Work!

Not the Babylonians, but only Euclid introduced a *definition* of the object: *concept* of "number".

> The definition is the *starting point* for mathematics as we know it.

The definition tells us *what is* the subject under discussion, *e.g.* "number" or "continuity". And this is *the very spot* at which the labour of mathematical thinking starts: to *define* the concepts. Mathematicians are not empirical scientists, who can rely on *sensual perception.*

> Mathematical thinking is not empirical, it does not consider the objects of our experience.

Which should those be?

Is a Mathematical Proof Beyond Reproach?

Another kind of mathematical labour is to *prove* a mathematical statement. This reverses the perspective. The mathematical proofs explore the capabilities of conceptual expressions.

A proof must be *strong,* at best *irrefutable.*

The most forceful proofs in mathematics are the proofs by contradiction. Mathematics *must* be free of contradictions. (Otherwise, *everything* could be proved: a theorem *together with* its negation.) However, not all mathematical proofs work by contradiction. Some proofs show: a definition is used *correctly.* Others demonstrate: so it *must* be, this is *sufficient!*

Mostly, the favourite proofs are *informal* and work without higher concepts. They are *not technical.* In case they are *completely* informal, they suffer from a severe shortcoming: they are *not necessarily* conclusive. In Chaps. 5 and 14 I shall present two such examples. Only a *formal,* technical proof ends the discussion and shows: there is no room for developing further arguments.

However, even the greatest did not always succeed and exhaust all the possibilities. In Chaps. 3 and 9, we shall see how two great thinkers reached the limits of their thinking, essentially, about the same issue. (We may wonder if they were aware of it.) As a further example of the change of mathematical knowledge we shall present a simple diagram, which could not have been drawn by Leibniz (or well as by his contemporaries and his predecessors), on p. 27.

From Confusion to Clarity

Mathematical thinking is formed within definitions and proofs. Through a plenitude of attempts, concepts are created, tested and rejected. This is how through creative thinking mathematics is produced. In the end, the so-called mathematical *knowledge* can be seen to be a conceptual *construction*—and as such it is always *open to discussion,* in its foundational aspects as well as in its concrete formulations.

The transformation of concepts is at the heart of the historical development of mathematics.

It is a fundamental object of this book to present this confusion of concepts as well as of proofs. And that not only in general, but in detail. Great mathematical thinkers (like Gottfried Wilhelm Leibniz, Johann Bernoulli, Leonhard Euler, Karl Weierstraß, Georg Cantor) were thoroughly convinced by their own mathematical arguments, which today's mathematics values to be not reliable. Some of them will be cited in their original form.

Growing Insight in the Formative Power of Definitions in Mathematics

Already, the early modern mathematical thinkers have been aware of the possibilities given by definitions or by mere creativity to shape mathematics (Chaps. 1 till 5). Only in the nineteenth century did a wider professional public take hold of these possibilities. Only then they agreed: geometry does not only exist in the plane but also, for instance, on the sphere—however, without validity of Pythagoras' Theorem, which only holds in the plane. Usually this is expressed as follows: there also exists a "non-Euclidean" geometry (which is in the true sense a "non-Pythagorean" one).

It took another century for the insight that such different configurations of a theory are not at all confined to geometry but are also possible for the most important and most influential mathematical theory, the Calculus, which we also call Analysis. A key-word therefore is Nonstandard-analysis. In this book it is shown: there also exists a *Cauchy an Analysis*—which is different from each of the theories known today.

The Change, Seen from a Philosophical Viewpoint

Having acquired this understanding of the formative power of human thinking, it is not difficult to take the trouble to assess mathematical thinking anew. This will require a *precise decomposition* of the *details of thought* in mathematics, which we examine.

We shall attempt a reassessment of the history of the most important branch of mathematics so far: of analysis.

And that in a way which focuses on the foundational ideas and concepts of the field. The decisive changes (or "revolutions") in the development of the theory will be shown. We shall not only stress the importance of naming the new but *also* point out which price is paid for any change. There is no innovation, which merely yields new results, considered to be "good", without the loss of something old. This is true also for the way of reasoning—and, consequently, for the development of mathematics as a whole. Usually, the relinquished parts are kept secret—not here. (An exception is the first chapter: the birth of modern mathematics. Even now, I am unable to express this huge transformation, this emergence of modern mathematics, in just a few words.)

This book concentrates on the (very) broad developments and the fundamental aspects of the relevant theoretical thought, the philosophical ones. My book *Die Analysis im Wandel und im Widerstreit*, which was published in 2015, is much more detailed (not least in mathematical aspects) and, consequently, more voluminous.

To sum up: I shall pay particular attention to the philosophical aspects of mathematical thought. However, the main focus will always be mathematics (analysis) and its actual development, not philosophical peculiarities and quibbles. It is the *actual* mathematics in the minds of the greatest mathematicians which is at stake here. This is presented by scrutinizing their original texts.

The *Formation* of Mathematics—or: The Transformations of Analysis

Analysis is the *most important* mathematical theory of modern times and thus—since the nineteenth century—also of the industrial era. It is a dominant part of this cultural and economic development and one of its *necessary* driving forces.

This book tries to trace *how* the central concepts of analysis in the last just 350 years were established: convergence and continuity, differential and integral, number and function. And it also offers some of the related proofs.

The Foundational Years

It starts with the premise of it all: the creation of the **formula** by Descartes (not by Viète!—see Chaps. 1 and 2). Because: without a formula there is no modern mathematics. (This has rarely ever been told.)

Already the very first mathematical formula contains an **"unknown"**. The next stage of development liquified it: the (discrete) "unknown" became the (continuous) **"variable"**. Newton as well as Leibniz had this idea (Chaps. 3 and 4). Only fairly recently it was established that Leibniz had created very precise mathematical concepts and arguments; for instance, he formulated a clear-cut **proof of convergence** and he invented the precise concept, which we call today **"Riemann integral"** (Chap. 3)—naturally in his own conceptual language.

Thereby, a philosophical abyss opened up: **do infinite numbers exist?** Two otherwise always fruitfully cooperating friends from the Champion's League of Mathematicians, Leibniz and Johann Bernoulli, were unable to find an agreement upon this question: see Chap. 5.

An Era of Pomposity: Algebraic Analysis

Following Johann Bernoulli, who transferred the focus of mathematical activity from geometry to calculations (or algebra, Chap. 6), his pupil Euler shaped the doctrine into **"Algebraic Analysis"**, which is ruled by the formula, now called **"function"**. In the Zeitgeist of absolutism of the eighteenth century, Euler institutionalized a formalistic etiquette, which disregarded some common-sense calculations (Chap. 7). D'Alembert and Lagrange worked on a further development along these lines. However, they did not end up with a genuine redesign (Chap. 8). Lagrange's attempt at a foundation failed because of a basic deficiency within his concepts.

The Implosion of Algebraic Analysis—and a First Attempt to Replace It

For the newly emerging practical perspective of the engineers, which became dominant in the nineteenth century, the formulae were of no use if they did not result in numbers. Bolzano as well as Cauchy did justice to this demand through being the first to move the **"value"** into the centre of the theory (Chaps. 9 and 10); Cauchy even succeeded in a complete theory. This gave analysis a new twist. Because nobody has yet described this transformation, I had to coin a new name for this doctrine: **"Value Analysis"**.

However, Value Analysis suffered from a difficult defect of omission: until then, nobody had invented a definition for a concept of "number", which was suitable for analysis—neither Leibniz, nor, especially, Euler. Ultimately, not even Bolzano did manage it, whereas Cauchy did not even try to do so. This **interregnum** lasted for about two generations (Chap. 11).

Implementation of a Capricious Value Analysis

Finally, Weierstraß (the last mathematician who dealt with a traditional form of analysis, Chap. 12), Cantor and Dedekind (Chap. 13) succeeded in cutting the Gordian knot. They swallowed the bitter pill and conferred the civic rights of mathematics to the actual infinite with the help of their **concepts of the real numbers.** Thereafter, an agreement on the true concept of function (from Riemann, not from Dirichlet!) was easily reached.

The ideas which are well known today are those of Cantor and Dedekind, for they were printed in 1872. **Weierstraß' concept of the real numbers** is still unknown and is presented here in print for the very first time in English. His idea was disclosed in the lecture notes of one of his students, which was discovered only in 2016; it is presented here in a short summary (Chap. 12).

Outlook: Axiomatics, Analysis Within Set-Theory and a New Kind of Formal Calculation

One generation later, Hilbert disavowed the traditional kind of mathematical thinking and paved the way for the structural mathematics à la Bourbaki of the twentieth century by his **axiomatization** of the concept of number (Chap. 13). At first, this attempt was not without flaws—even a Hilbert can make a mistake (p. 206).

The transformation of Value Analysis into **Set Analysis** in the twentieth century is not a subject of this book, but the basic philosophical problem of this approach is clearly indicated here (p. 199). This would quickly run into technical details, and besides, the result can be found in any modern textbook. But for the sake of clarity, we emphasize

Set Analysis only arose in the twentieth century

As a note to the specialists: Georg Cantor defines *each* converging sequence ("Cantor sequence") of rational numbers to be a real number (p. 191). (He defined two "numbers" of this kind to be "equal", if they differ by an "infinitely small quantity", which is a sequence converging on the limit zero.) Cantor did not create the concept of "equivalence relation" (see pp. 62, 120), which also originates in the twentieth century. He created set-theory only two decades after his definition of a real number. It is very astonishing (and only known since 2017): the first and only mathematician who fruitfully used set-theoretical constructions to create a concept for the real numbers in the nineteenth century was Karl Weierstraß (Chap. 12)—besides, almost without being noticed, Traugott Müller and François Bertrand, and, of course, somewhat overbearingly: Richard Dedekind (Chap. 13).

Finally, in Chap. 14, I shall reveal a little-known branch of analysis, which today is usually hidden behind logic or axiomatics. Here, my starting point is another unpublished manuscript, for once a notebook from the splendid calculator Curt Schmieden dating back to the early 1950s. Schmieden did two things. First, he changed the concept of the real numbers. Secondly, he transformed the way *of calculation with the infinite:* instead of describing the infinite in the formulae only (a) at the *start* and (b) thereafter only by " ...", he stuck to the demand (c) to be precise also at *the end of the formulae* by saying *explicitly* what one means by the infinite there. Schmieden demanded **a precise end for the infinite.** (Consequently, his formulae are longer than the usual ones, and, moreover, they require twice the usual effort—because of the necessity to think about their ending, not only about their start. However, these efforts always stay *elementary.*) With the help of Detlef Laugwitz, Schmieden's ideas became the first impulse for the invention of a new theory, which nowadays is called **"Nonstandard-analysis",** in 1958. However, in my opinion, the potential of Schmieden's ideas is far from being exhausted.

The First Mathematical News in This Book: The Archetype of Today's Analysis (from Cauchy)

The central concepts of modern analysis, "convergence" and "continuity", were formulated crystal-clearly by Bolzano and Cauchy. However, it was just Cauchy who developed the theory *completely*. He based it upon the old basic concept of "function" (to be a formula) and the new concept of "value of the function" and added the precise notions of "derivative" and "integral".

Historically, Cauchy's notion of "value of a function" was absolutely new. This mathematical object had never been defined before. Surprisingly to us, Cauchy thought about this concept rather differently from the way we do (p. 124).

This *different* understanding of the concept of "value of the function" necessarily implies consequences for those concepts, which *derive from* it. These are especially the notions of "convergence", "continuity" and "derivative". They got meanings which differ from the meanings of their modern counterparts. (Cauchy's definitions of "convergence", "continuity" and "derivative" *read* just like ours, but their *meaning* is different!) "Different" implies: they make some *other* theorems true, not those we have today.

In Chap. 10, this is shown in detail. To sum up: the *archetype* of analysis was different (and it was also more easy) from its modern development with its changing concepts of "function" and "value of the function". (Seen from this perspective, Cauchy's concepts may not appear as such a surprise?)

The Second Mathematical News in This Book: A Third Construction of the Real Numbers (by Weierstraß)

Besides the well-known sources of the two traditional concepts of the "real numbers" (it is true: I add a hitherto unknown early predecessor of Dedekind, from 1838!), I present a completely new one: from a manuscript which was found by chance in the mathematical library of the university Frankfurt am Main as recently as 2016. This precise construction of the concept of real numbers is the most elementary one known today and has never been re-invented by others.[1]

The fact that Weierstraß lectured on it during and after Cantor's and Dedekind's ideas were already published might show that Weierstraß thought his own idea to be superior to the other two.

It is my opinion that Weierstraß' construction of the real numbers prospectively should really replace the others, because it is the most elementary one. Cantor's construction uses the concept of convergence—which is a *petitio principii* within

---

[1] Detlef D. Spalt 2022, *Die Grundlegung der Analysis durch Karl Weierstraß – eine bislang unbekannte Konstruktion der natürlichen und der reellen Zahlen.* (Springer).

the scope of a deductive composition of analysis. And compared to Dedekind's (actually: Traugott Müller's!) idea, Weierstraß' is more elementary, because it does not rely on the puzzling concept of "least upper bound".

No gain without pain. The said passages (in Chap. 12) are *very* elementary mathematics, but as they are yet unknown, they are completely new territory for *any* reader—and understanding something new always requires much more effort than the recognition of something already known. Thus, even mathematicians will have to read these passages with some more care than others. Mathematics *is not* trivial. That is why these subsections go into some details.

The Historiographical Hallmarks of This Book

In Substance

It is the maxim of this book to take the former mathematics *in its own rights* and to present the mathematical thinking of previous mathematicians *as it was*— instead of *transforming* it into that caricature, which results from its *translation into modern language and modern thought.* Only in this way we are able to do justice to history, because the earlier facts existed without the later ones and without our actual thinking—just as nowadays we do not have the faintest idea about the shape of mathematics of the next or even the following century. Only in case of Weierstraß' concept of number, I make an exception as a concession to the reader: to ease their understanding.

This attitude is quite unusual, and it has to be implemented especially in Chaps. 10 and 11, i.e. when we look at the nineteenth century. For this is a period in which mathematicians, who are not historians, sometimes are led astray. This is because the "Value Analysis" of that time appears to have much similarity with today's set-theoretical analysis. The historical layman as well as the historical laywoman feels compelled to get along with those texts: *look, these are the same concepts as ours!*

Those who are not aware of the fact that mathematical thinking underwent transformations *over time* and also that *mathematical concepts are invented* (or who is not at least willing to think that this *could* be the case) will assume that the former concepts are our modern ones and subsequently will often reach wrong conclusions. They will take Cauchy's analysis to be identical with today's—which definitely is a *mathematical error.* It was just its *archetypal form.*

Unfortunately, I myself got into this trap when I worked for the first time on original texts between 1977 and 1986. However, not only mathematics changes, but also—at least sometimes— one's own understanding of the world. Some succeed in learning new things.

During the last half century, all this led to some grotesque misunderstandings of the analysis of the nineteenth century. The most dramatic example thereof (mathematically as well as historically) will be shown in Chap. 10 in some detail, another one in Chap. 11. At the beginning of Chap. 11, a comprehensive presentation of my historiographical method is given, which is an example of the complex facts stated in Chap. 10: the historiographical judgement of thoughts and concepts within analysis which *differs* from *all* our actual theories—but which is, nevertheless, *consistent in itself* (just as the Cauchy ian Analysis). History of mathematics is not only able to *detect* (seen from today) single new ideas (like Weierstraß' concept of numbers) but even an entire new *theory!* (Well, of course one has to reckon with that …)

For that reason, Chap. 10 is of a somewhat more complicated structure than the others. There is not only the subject presented (*i.e.*: how did Cauchy think analysis?) but also the *two* (!) mathematical misconceptions that Cauchy's thinking will be unravelled—*together with* the identification of the errors involved.

In Method

This book is solely based on original sources. (All of them are given in English, sometimes the translations are new.[2]) Normally, these texts are well known or, at least, easily accessible (in our digital world far more easily then ever). In Germany, the digitalization is sadly behind many other countries: for instance the French one, with its portal *Gallica*, or the Swiss one, with the portal www.e-rara.ch which presents the noblest sources. Of general help was also the Californian *Internet Archive;* however, things seem to have changed.

All Told

This book aims to ease access to the history of analysis. It gives an overview of the important lines of its development: particularly by scrutinizing the definitions of the basic notions, followed by some detailed discussion.

Mathematics changes—like all the other kinds of human thought. "Changes" means: it becomes *different*. This becoming different is the subject of history of mathematics—whereas the actual mathematical thinking usually strives not for a difference, but, instead, for more.

---

[2] My German book *Die Analysis im Wandel und im Widerstreit* has, besides the German wording, also the original.

Chapter 1
The Invention of the Mathematical Formula

Today, mathematics is unthinkable without formulae. Without:

$$(a + b)^2 = (a + b)(a + b) = a^2 + 2ab + b^2$$

nothing works at all.

Did formulae always exist? Or did somebody invent them? Of course, somebody *must* have invented them, they cannot just appear from nowhere!

Then, who did invent the mathematical formula? How can anybody have such an idea? And why?

These are the three questions, which will be addressed in this chapter.

Who Invented the Mathematical Formula?

Strangely, in school we learn many mathematical formulae, but not where and when they were created.

It is even more odd that this is nowhere written down! There is no encyclopedia, including Wikipedia (at least not until today, 2021), that mentions who first thought of the mathematical formula.

However, reading books can educate and here is the answer to that question:

> *The mathematical formula was invented by René Descartes.*

René Descartes—who is he and does one need to know him?

Yes, one should know René Descartes, at least a little bit! He lived from 1596 until 1650. Being French, as is obvious from his name, he nevertheless lived many years in the Netherlands. In the twenty years between 1629 and 1649 he moved about 24 times in order to avoid being disturbed by the authorities.

© The Author(s), under exclusive license to Springer Nature Switzerland AG 2022
D. D. Spalt, *A Brief History of Analysis*,
https://doi.org/10.1007/978-3-031-00650-0_1

René Descartes thought in ways unacceptable to the officials. The Churches of central Europe were part of the ruling elite which possessed secular power. It was, for instance, the Catholic Church that sentenced Galileo Galilei in 1613 for his claim that the Earth is moving and that the Sun is at the centre of the cosmos. When Galilei repeated his belief in 1613, he was forced, under threat of torture, to revoke his announcement and incarcerated by the clerical officials. During that time the Thirty Years' War (1618–48) was waged. What begun as a war about religion had turned into a war solely about power and influence.

Descartes did not want to share Galilei's destiny. Thus, he preferred not to publish certain texts. In August 1634, Descartes managed to view Galilei's forbidden text, in secrecy, for a few hours. There he read what he could not publish.

Come to that, the text contained no mathematical formulae. Galilei did not know any. For Galilei it was circles, triangles, and other geometrical objects that according to him constituted the "letters" of mathematics. These kind of "letters" had already existed for two thousand years—Descartes radically changed all this.

How Did Descartes Invent the Mathematical Formula?

Descartes, too, is a child of his time. As with every thinker, his thought is shaped by his experience. This experience was the increasing importance of individual vocation in order to gain social status, as well as the mechanization of the world. Independently of their status at birth, an ever increasing number of artisans, artists, and engineers shaped their lives through their professional activities. Wearable watches, which were, however, not very precise, had existed already for about a hundred years ("Nuremberg Egg"); pumps became important in order to provide big cities, such as London and Paris, with enough water. Fortresses were built, and so on.

It was Descartes' special strength to express his Zeitgeist by means of *concepts*. He thus anchored philosophical thought in the thinking of the individual: "It is thinking, thinking alone, which cannot be separated from me: I am, I exist, this is certain". This "I" is of course not the man Descartes but the abstract encapsulation of the concrete individual: I forge iron, therefore I can live (from it)! It is not the social status at birth that determines a life, but one's own doing. Descartes subsequently becomes the founder of modern philosophy, a philosophy that fitted the new, the bourgeois and mechanized world of the people. And this without the backing of the authorities. This is a revolution.

And how does Descartes set the foundation for modern mathematics? There are two aspects which act in unison.

The first aspect is the *method*. Somewhat pedantically, Descartes demands: Focus your penetrating mind on the simplest things! Disregard everything superfluous and unnecessary! Represent everything by clearly perceptible figures. Keep your notations as brief as possible!

Decisive for all this was Descartes' basic conviction: everything which can be grasped "clearly and distinctly" is also *true*. And *only* that is true.

The second aspect is this: only Geometry is *real* mathematics. Thus: numbers and calculations are only then mathematics if they are *transferred* into Geometry.

The reasons behind Descartes' thinking will be discussed later. First of all, the focus should be on how Descartes came up with the mathematical formula.

Transfer Arithmetic into Geometry!

So if *real* mathematics is restricted to Geometry, it will be necessary to shift the numbers into Geometry. And, of course, the calculations as well.

To transfer the numbers into Geometry is simple: The number 1 becomes the *length* 1, the number 2 the … So, instead of mere *numbers* (as it is with arithmetic) we have *straight lines* of the relevant lengths: geometrical objects.

We still need to deal with the calculations, which must happen with *straight lines*.

Adding and subtracting is simple: the straight lines are put together or they are taken away from each other. (Of course, *negative* numbers do not exist: what should that be, a "negative" number?)

There are still more complex forms of calculations: multiplying and dividing.

This proved problematic for Descartes. First, he did the obvious and declared: the product of two lines is an area: "If we multiply _ a _ by ___ b ___ we put them together at right angles in the following way ___ b ___ in order to get the rectangle

a

b

a .

However, if one wanted to multiply the result by c, one would have to imagine

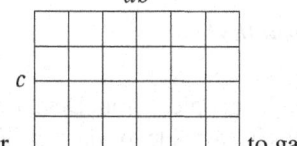

ab

c

ab as a line a b _____ ab _____ in order ___ to gain abc".

These were Descartes' initial thoughts, but it becomes immediately evident: this is *not* the way to do it! An area is not a straight line but, according to Descartes' idea, the product ab is first an area but then, after multiplication by c, the *same* product ab suddenly is a straight line.

It is, therefore, justified that Descartes interrupts his uncompleted text at that point—which he really did.

(He himself did not gave a reason for it. One can only guess. Yet, this is simple considering what we just have said.)

About ten years later, Descartes presented the solution to his problem. In the only book on mathematics ever written by him (and which was of course called *The*

Geometry), he explains multiplication completely differently:

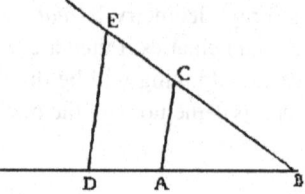

"Let, for instance, AB be the unity. In order to multiply BD by BC, I only have to join the points A and C, then to draw DE as a parallel to CA and BE will be the product of this multiplication".

Descartes has *two* new ideas: First, fix a "unity" in a completely arbitrary way. Secondly, use the *principle of similar triangles*. Then all becomes simple. If $AB = 1$, $BD = a$ and $BC = b$, then all one needs to do is to connect A and C and to draw the parallel to AC through D—finished! The intersection gives us point E and the line segment BE. According to the principle of similarity we have:

$$\frac{BD}{1} = \frac{BE}{BC} \quad resp. \quad \frac{a}{1} = \frac{BE}{b}, \quad \text{therefore} \quad ab = 1 \cdot BE = BE.$$

Knowing the principle of similarity, it is simple. It was known to mathematicians at least since it was mentioned by Euclid, about two thousand years ago.

This is all: thus Arithmetic is transferred into Geometry, just as Descartes had wished. (Division is easy, the principle of similarity is used slightly differently.) From this point on, instead of *calculating,* one has to *draw!*

Now we still miss the formulae! They appear, at least by Descartes, all by themselves.

Solve Problems!

Let us read a longer piece from Descartes' *Geometry,* it is worth doing! Remember that Descartes is now able to *calculate with straight lines.*

> If, then, we wish to solve any problem, we first suppose the solution already effected, and give names to all the lines that seem needful for its construction—to those that are unknown as well as to those that are known. Then, making no distinction between known and unknown lines, we must unravel the difficulty in any way that shows most naturally the relations between these lines, until we find it possible to express a single quantity in two ways. This will constitute an equation, since the terms of one of these two equations are together equal to the terms of the other. We must find as many such equations as there are supposed to be unknown lines.

Incidentally, since antiquity, this procedure has been called "Analysis" (in opposition to Synthesis), leading to today's term "analysis". Descartes' text carries on for a few more sentences and then he writes some formulae:

I may express this as follows:

$$z \, \infty \, b, \quad \text{or}$$

$$z^2 \, \infty \, -az + b^2, \quad \text{or}$$

$$z^3 \, \infty \, +az^2 + b^2z - c^3, \quad \text{or}$$

$$z^4 \, \infty \, az^3 - c^3z + d^4, \quad resp.$$

That is, z, which I take for the unknown quantity, is equal to b; or, the square of z is equal to the square of b diminished by a multiplied by z; or, the cube of z is equal to a multiplied by the square of z plus the square of b multiplied by z, diminished by the cube of c; and similarly for the others.

> These are the first original equations given as formulae in mathematics.

(To be precise: not entirely original. Descartes used a different equality sign than that which is used today. But this is not important. It is decisive that a *sign* is a *symbol* that is used instead of a *word,* as it was the case.) But the formulae do not depend on language. Formulae are a *new language:* the language of the new mathematics.

What is so exciting about it?

Yet, Descartes did not attend any school where he could have learnt this. His contemporaries wrote such things completely differently.

For instance, François Viète (1540–1603). Viète is officially named as the "founder of algebra", by which one really says that he invented *formulaic expressions.* But he did not. Because in 1593, Viète at most comes up with such things as this:

$$\frac{B \text{ times } A}{D} + \frac{\begin{array}{c} B \text{ times } A \\ -B \text{ times } H \end{array}}{F} \text{ are said to be equal to } B.$$

Interestingly, Viète puts *vowels* in place of unknowns, while Descartes is using the *last letters* of the alphabet. (The vowels represent for Viète the unknowns— in similarity to the inventors of the letter-based writing system, the Phoenicians, who only used consonants and who considered vowels to be "unknown letters"; in Arabic too, vowels were never written, till today. Only the Greeks introduced letters to represent vowels. But this observation by Otto Hamborg is only added in parentheses.)

However, Viète often writes equations completely differently, for instance, like this:

Given be B times A to the square plus D planar times A equals Z corporal.

This looks fairly awful. A careful *translation,* after some thought, could be written as follows:

$$B \text{ times } A^2 + D^{\text{planar}} \cdot A \text{ is equal } Z^{\text{corporal}}$$

Descartes would simply have written:

$$az^2 + bz = c$$

Descartes would completely have omitted the denotations "planar" and "corporal". Yet, Viète uses those notations. He thus shows what *kind* of object he is considering. In other words: for Viète it is not just the letter that denotes the object in question, but the *meaning* of the letter is determined even further. That means, for Viète, letters are not *general* symbols but are denoting, together with their labels, *specific* kinds of objects.

The last point is of *ideological* importance: Viète put particular importance on the "Law of Homogeneity". That means all the objects which are to be added (or subtracted) *must* be of the same kind, i.e. lines, areas or solids. For, according to Viète, it is impossible to add something "one-dimensional" to something "two-dimensional". And we remember: Descartes thought *initially* the same! But only in the beginning.

The mature Descartes can think very differently. Because the mature Descartes— we remember his application of the principle of similarity!—has chosen a "unity". Descartes can *insert* this unity (let us call it "1") whenever he likes to do so *as a factor or several factors.* Therefore, he can read the last equation thus:

$$az^2 + bz \cdot 1 = c \cdot 1 \cdot 1$$

or also like:

$$\frac{az^2}{1} + bz = c \cdot 1$$

and now it is "homogeneous"! (Of course, he could write something completely differently, such as:

$$az^2 \cdot 1 + \frac{bz}{1 \cdot 1} = c \cdot 1 .$$

It makes no real sense, but it is not wrong!)

The later Descartes has shown: *the Law of Homogeneity is superfluous.* It is an unnecessary boundary for mathematical thought. More precisely: for calculations.

Descartes managed to do this by *defining calculation in a new way.* He started by fixing a "unity", which makes *calculations with line segments easy.*

Viète was not able to do this. Because of this he had to fall back on additional attributes such as "planar" and "corporal" besides his letters.

We learn: that which is for one person "the first and generally valid Law of Equations", is, *if one thinks about matters completely differently,* entirely superfluous. In this case, the matter is calculating.

> *The fixing of a unity makes the Law of Homogeneity superfluous.*

Thus, what is allegedly a law of thinking can be *proven* to be obsolete by means of a *different form of thought. If one thinks differently, one's judgement might change too.* This seems self-evident and *applies to mathematics too.* Mathematical judgements too (propositions and theorems) are based on initial rules. They can be changed by *thinking differently* or *something different.* One needs, of course, an idea how this could be done. (As we saw with Descartes, and as we will see yet again, such ideas are only found with difficulty.)!

Descartes did have such an idea: the arbitrary fixing of a unity and the application of the principle of similarity. Strangely, after all: Descartes' first equations are *homogeneous!* Is he following old habits?

And as we are already penetrating fairly deeply into this matter: *By utilizing the principle of similarity, Descartes assumes the validity of Euclidian mathematics.* He does this without spelling it out. Maybe he was not even aware of it. Because the first sentence in his textbook for mathematics says: "Any problem in Geometry can easily be reduced to such terms that a knowledge of the lengths of certain straight lines is sufficient for its construction". There is no real justification for the self- congratulatory tone of the statement: *without giving it a mentioning,* Descartes assumes Euclidian theory.

Considering all this, it appears to be completely nonsensical to designate Viète as the inventor of the mathematical formula. It is Descartes, who clearly deserves this honour. His idea—it took him years of thinking in order to come up with it—finally looks simple: in order to solve a problem, only write down the relevant straight lines (numbers), *by means of* SINGLE *letters,* as well as the arithmetical operations which connect them. For the known quantities use the first letters of the alphabet, for the unknowns the last. That this is *not really easy,* as can be seen in school: many have problems in grasping this idea.

Of course, this does not mean that Viète's partly formalized way of notating was entirely useless. Quite the opposite: Viète was the first who was able to write down how to get the solution of an existing equation in an orderly and succinct way. Viète also gives examples of how to calculate the coefficients (which he did not call so) of an equation from its solutions—a procedure which we now refer to as "Viète's formulae".

Why Does Descartes Have Those Ideas?

Why does Descartes only consider Geometry to be proper mathematics, but not Calculating (Arithmetic)? The answer is deeply anchored in Descartes' world view.

We know already that Descartes kept away from the authorities as much as possible. He knew: if his *real* thoughts became known, he would be persecuted like Galilei. So, what were Descartes' revolutionary thoughts about the world?

These: a human being is not what is usually claimed, a combination of body and mind. Instead, body and mind are two *fundamentally different* and, therefore, separate entities. More precisely: there are only two "substances" which exist in the world. One is *matter* and one is *mind*; or in a different formulation, *extension* and *thought*. (That is why one usually speaks about Cartesian *dualism:* for Descartes there exist only *two* fundamental substances.)

All things in the world are mere *appearances* of those two substances. Solids and space are only "modi" (the plural of the Latin word "modus") of matter (or extension); soul and time (namely: memory) modi of the mind (or of thought).

Strong words, which are of course heretical: God merely a modus of thought?

But this was how Descartes thought. He also wrote it down. Yet, in order to avoid trouble, he could not express his thoughts so clearly. For that reason, his philosophical texts contain inconsistent statements: safety measures. In order to understand Descartes, one needs to read his texts very carefully while considering his political situation.

Thus, according to Descartes, there are exactly two substances: matter (or extension) and mind (or thought) with their exemplifications of body and soul.

Bodies or *solids,* in being modi of matter are appearances caused by motion. *Souls* are modi of mind. Subsequently, body and soul are *really* distinct; a unity of both does not exist. *Not really* distinct are, however, matter and bodies; thought and mind; your soul and my soul; soul and God.

But what are "numbers" for Descartes? Obviously they are *not bodies*. Although we *count* bodies, the numbers we *deduce* in the process, are not independently existing entities. "Numbers" derive from counted objects and thus are *connected* to them. If one disregards these connections one only arrives at "nonsense" (Descartes might also have thought of numerological practices). Instead, one has to "abstract" from the "numbers".

If a right-angled triangle has shorter sides of lengths 3 and 4, the arithmetician says that the third side has length 5. The mathematician, that means the geometrician, however, will say that it has the length $\sqrt{a^2 + b^2}$ (if a and b are the shorter sides): "The two parts a^2 and b^2 remain distinct, while they are mixed up in the number 5".

The number is distinguished from the counted objects in the same way as the quantity from the extension. Therefore, there cannot be a science of numbers (no arithmetic)—but definitely one of extension (Geometry).

Descartes, therefore, *had* to transfer Arithmetic into Geometry where numbers became (as a matter of simplification) straight line segments. But, as a consequence, one had to calculate with lines. Therefore, it was important to pay attention to the correct method: "Direct your attention toward the simplest objects! Disregard everything superfluous and unnecessary! Express everything by figures of the intuition! Keep your notations as short as possible!"

The result reached by Descartes was *the purely symbolic mathematical formula.* But the formula is the decisive means of construction for modern mathematics. Nobody before Descartes had this idea.

Six years before the publication of Descartes' *Geometry,* with its purely symbolic formulae, another book containing such *formulae* was published. It was found among the writings of Thomas Harriot, who had died ten years earlier in 1621. (Harriot even wrote down the equality sign which is used until today and which was introduced two generations earlier in England.) Admittedly, the author did not use the symbolism of powers, which often led to confusing and prolonged calculations. Besides, all the equations follow the Law of Homogeneity—clearly, because Harriot did not have the same revolutionary idea as Descartes.

Descartes always denied having had knowledge of this book before 1637 and such knowledge could not be proven until today.

It became quickly apparent that mathematics had gained a completely new perspective for further development. A case in point is that a purely symbolic equation becomes *a new mathematical object*—for which one can create a theory. One can, for instance, carry out calculations with them.

What Is x for Descartes?

Clearly, Descartes' equations deal with "numbers". The known numbers are denoted by the letters at the beginning of the alphabet; others are unknowns: those marked by "x". Initially, Descartes rather writes "z", where we, today, usually write "x", yet, in the last chapter he changes to "x".

To *calculate* an unknown x may be simple:

$$x - 1 = 0$$

has obviously the value 1.

Sometimes, there are *several* values. In the last chapter of his book, Descartes treats equations as the objects of calculations:

If one multiplies the two equations,

$$x - 2 = 0 \quad \text{and} \quad x - 3 = 0$$

one gets

$$x^2 - 5x + 6 = 0$$

or

$$x^2 = 5x - 6,$$

and this is an equation, where the quantity x can take the value 2 as well as the value 3.

Thus, x is *not necessarily uniquely determined.* Of course, there could be even more values which are represented by x; one merely has to multiply more than two of those equations.

Of course, Descartes has no "negative" numbers—because his "numbers" are *line segments.* If an equation leads to such a result, as in

$$x + 5 = 0,$$

then it has a "false solution" which is 5.

The result:

| *Descartes' "x" denotes one or more* UNKNOWN *numbers.* |

Descartes did not need to say what *"numbers" are:* he had transferred them into *lines.* The meaning of *line segments* ("lengths") could already be found in Euclid. He also holds that there are "rational" and "irrational" (for Euclid "incommensurable") numbers.

Albeit, until the concept of "irrational" number was finally established another 212 years would pass. We will come to this quite toward the end, in Chaps. 12 and 13.

Literature

Bos, H. J. M. (2001). *Redefining geometrical exactness. Descartes' transformation of the early modern concept of construction.* Berlin: Springer.

Cajori, F. (1929). Controversies on mathematics between Wallis, Hobbes and Barrow. *The Mathematics Teacher, 22,* 146–151.

Descartes, R. (1637a). *Discours de la methode.* Leyde: Ian Maire. http://gallica.bnf.fr/ark:/12148/btv1b86069594.r=descartes+discours.langDE.

Descartes, R. (1637b). *Geometrie.* Darmstadt:Wissenschaftliche Buchgesellschaft, 1981. reprint of the 2nd ed. from 1923, Leipzig. German by Ludwig Schlesinger 1894.

Hobbes, T. (1662). Seven philosophical problems and two propositions of geometry. In W. Molesworth (Ed.) *The english works of Thomas Hobbes* (Vol. VII). London: John Bolton, 1845. Second reprint: Aalen: Scientia Verlag, 1966.

Irrlitz, G. (1980). *René Descartes – Ausgewählte Schriften.* (Leipzig: Philipp Reclam jun.)

Schmitz, M. (2010). *Analysis – eine Heuristik wissenschaftlicher Erkenntnis.* Freiburg im Breisgau: Verlag Karl Alber.

Smith, D. E., & Latham, M. L. (Eds.). (1954). *The geometry of René Descartes.* New York: Dover.

Stevin, S. (1586). *De Beghinselen der Weeghconst.* Leyden: Christoffel Plantijn. http://www.google.com/books?id=_wo8AAAAcAAJ&hl=de.

Chapter 2
Numbers, Line Segments, Points—But No Curved Lines

Mathematics Is in Need of Systematization

In 1637 Descartes invented the pure symbolic formula for mathematics. This became possible only after having unmasked the Law of Homogeneity as null and void, as an unnecessary curtailment of thinking.

Only in the course of time is the *quality* of an idea revealed. Often it is the *novelty* of an idea that hinders its appreciation and its dissemination. Not everybody loves news; clinging to habits is more conducive to comfort.

Thomas Hobbes (1588–1679) was a contemporary of Descartes, being born eight years earlier and dying 29 years after him. The famous *Encyclopedia Britannica* describes him as a philosopher and a political theorist but Hobbes considered himself also to be a competent mathematician. Nonetheless, until today he is not known for this quality to historians of mathematics. This is generally agreed and will not be questioned here.

In this way, Hobbes may be a good example of a thinker who is highly esteemed in some matters (like philosophy and politics) but cuts a poor figure in other subjects (mathematics).

Considering a calculation like $9 \cdot \sqrt{2} = \sqrt{9^2 \cdot 2} = \sqrt{81 \cdot 2} = \sqrt{162}$, he asks, in the year 1662, how this could be possible—for the first product is a rectangle but the second a line segment.

> I see the calculation in numbers is right, though false in lines. The reason whereof can be no other than some difference between multiplying numbers into lines or planes, and multiplying lines into the same lines or planes.

Hobbes was unable to reason without the Law of Homogeneity in the year 1662, *although* Descartes had uncovered it as a superfluous barrier for reasoning in 1637.

© The Author(s), under exclusive license to Springer Nature Switzerland AG 2022
D. D. Spalt, *A Brief History of Analysis*,
https://doi.org/10.1007/978-3-031-00650-0_2

True and False Roots

Descartes supplied mathematics with the new entity: the pure formal equation. And what do mathematicians prefer? They prefer to calculate. If there exist new entities, they calculate or attempt to calculate with them.

And so did Descartes. At first he multiplied his new objects, the equations. The equation

$$x - 2 = 0$$

is very simple. Its solution is natural: $x = 2$. Descartes adheres to the traditional name for the solutions of an equation and says: 2 is the "root" of this equation.

The equation

$$x - 3 = 0$$

has root 3. And now it happens: the equations are manipulated; they are *multiplied*. We have already seen this in the preceding chapter. The *product* of these two equations is:

$$(x - 2) \cdot (x - 3) = x^2 - 5x + 6 = 0.$$

The left side directly shows that the *two* roots are 2 and 3, for a product $a \cdot b$ is zero, if at least one of the two factors is 0.

What Are False Roots? And What Is Their Use?

Now, instead of $x - 3 = 0$ let us choose the equation

$$x + 3 = 0.$$

What is the root of this equation? Today we say: the root is -3. But Descartes did not! Instead he says: this equation has the "false" root 3.

Why?

Well, he has transferred arithmetic to geometry and taken numbers to be line segments. To consider 3 as a line segment seems easy if you have a unity, as previously proposed by Descartes. However, what about -3? What kind of *line segment* could this be?

Today, mathematicians, physicists, or engineers will answer immediately: "the opposed line segment!"

That is what we have been taught: a straight line has a direction. However, this is *anything but* clear. A straight line does not have a *direction* by *itself*. (Maybe it has two of them, but surely not *one*. Indeed, the idea of the directions of a straight line had been proposed by the Dutch engineer Simon Stevin (1548/9–1620) in the year 1586, but in 1637 the philosopher Descartes was not interested in this proposal.)

Who uses "opposed to" has to assign a direction to *all* straight lines. He or she only knows *directed lines* as opposed to *straight lines* per se.

Descartes did not see the need for this distinction. Why should he have done? What is the use of straight lines having a direction in geometry? None at all! Euclid did not know straight lines with a direction. And as we will see, Leibniz, two generations later, did not subscribe to the idea that straight lines have a direction in geometry. But without "directed" lines there can be no "opposed" line segments! It is only calculation *with* line segments that needs a direction, not the calculation *of* line segments. This was shown by Descartes.

When we think, in various ways we presume things *without* being aware of them, for example, that straight lines *always* have a "direction". Obviously, this view is possible. But as Descartes shows, it is far from necessary to think in this manner.

> "Negative" line segments do not exist in classical geometry and neither in Descartes' view.

Consequently, the equation $x + 3 = 0$ does not have the "negative" root -3 but the "false" root 3. And why should it?

If we now look at the product

$$(x - 2) \cdot (x + 3) = x^2 + x - 6 = 0,$$

it has the ("true") root 2 and the "false" root 3.

Turn False into True

Of course, the "false" roots are an eyesore.

Descartes has two ideas on how to get rid of them.

1. *Swap the signs of the terms with x, x^3, x^5, etc.*
   That means, instead of

$$x^2 + x - 6 = 0$$

   you consider

$$x^2 - x - 6 = 0.$$

Because we have $x^2 - x - 6 = (x + 2) \cdot (x - 3)$, this equation has the ("true") root 3 but also the "false" root 2. Easy come, easy go. What turns false roots into true ones, equally turns true roots into false.

Can't we do anything about that? Yes, we can:

2. *Increase the roots by a sufficiently large quantity.*
   Consequently, we enlarge the roots of our initial equation

$$(x - 2) \cdot (x + 3) = x^2 + x - 6 = 0$$

with the "true" root 2 and the "false" root 3 by, let us say, 4:

$$y = x + 4 \,.$$

Then we have to do some calculation:

$$(x - 2) \cdot (x + 3) = ([y - 4] - 2) \cdot ([y - 4] + 3)$$
$$= \quad (y - 6) \quad \cdot \quad (y - 1) \quad = y^2 - 7y + 6 = 0 \,,$$

and indeed we now get two "true" roots for our equation: 6 and 1.

This approach always succeeds: we are always able to find a sufficiently large number which, if added to all roots of the equation, gives *only* "true" roots of it. In this way we do not need to encounter the "false" roots. We may get rid of them.

The Geometrical Advantage of Equations

In his *Geometrie,* Descartes has the proud boast that he is able to solve a problem that defeated the ancient mathematicians. This is right, but the problem is complicated to such a degree that I cannot present it here. We have to confine ourselves to the fundamentals of the case.

Analysis: From Problem to Equation

Descartes describes how he tackles the problem (all *emphases* are added):

> First, I suppose the thing done, and since so many lines are confusing, I may simplify matters by considering *one of the given* lines and *one of those to be drawn* (as, for example, AB and BC) as the *principal lines,* to which I shall try to refer all others.
>   Call the segment of the line AB between A and B, x, and call BC, y. [...]

We note two important aspects: the first is Descartes' *method.* We have already seen this in Chap. 1 (p. 4): *Consider the problem to be solved, name the quantities and look for relations between them.* From antiquity this method has been called "Analysis" (Fig. 2.1).

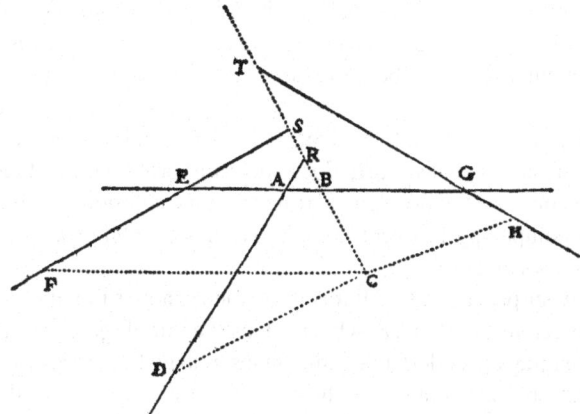

Fig. 2.1 The Pappos Problem, or: Descartes' system of coordinates (*Discours* 1637, pp. 309, 311)

There is a second aspect: Descartes introduces something, which we later, after its full development, call "coordinate axes". He has "principal lines" AB and BC and calls the "points" on AB "x", those on BC "y".

So much to consider!

1. At once we are accustomed to the "axis" AB being perpendicular to BC. But for Descartes things were different: the angle between the axes was arbitrary.
2. For Descartes, "x" and "y" were no longer "unknowns" (unknown *numbers*) but "arbitrary" *points*.
3. The intersection of the axis, B, is *not the "origin"* of the coordinate system and does not signify a *"zero"* for those axes. This is due to the fact (which is not shown here) that Descartes only works with *positive* lengths. Naturally, he *adds* and *subtracts* these *positive* line segments (or at least *almost* always).

We simplify the matter and avoid Descartes' tedious path to solve his problem. We merely examine his solution. It is to be found in the second chapter of *The Geometry:*

$$y = m - \frac{n}{z}x + \sqrt{m^2 + ox - \frac{p}{m}x^2}\,,$$

where m, n, o, p, and even z (!) are fixed quantities, which he introduced during his Analysis.

Please note: this equation is *homogeneous*. Descartes did not write it in the elegant way shown above but used the old notation: "mm" instead of "m^2", etc.

He gives the following explanation for his equation:

This must give the length of the line BC, leaving AB or x undetermined.

Thus did Descartes reach the goal of his Analysis of the problem: he found the equation which shows the connection between the "indeterminates" x and y. In respect of calculating, or to be more sophisticated, algebraically, he *solved* the problem.

But according to the classical view, this is not sufficient: it remains to be *proved* that the result of the Analysis *really* does solve the problem. In other words, one must be able to compose the problem starting from the solution. From ancient times this process of composing has been called "Synthesis". Only the Synthesis shows that the Analysis was *right*.

This is hardly surprising: after all this is geometry, not arithmetic! Descartes still has to show *the geometrical object* which solves the initial (*geometrical!*) problem.

Our understanding as modern mathematicians who no longer equate mathematics to geometry but are accustomed to algebra, can easily appreciate that the above equation *solves* the problem. Needless to say, for Descartes and his predecessors things were very different.

Interjection: Continuity

Descartes still has to draw the *line* representing the arbitrary quantities which are essential to the equation. There is a categorical distinction between *lines* and *points:* lines are *continuous*, but points are not.

That lines are *continuous* objects means that they cannot cross without intersecting.

In the case of two *lines of points* this is not guaranteed: maybe one of them—or even both!—does not have a point where they cross, albeit *without meeting*.

This is the classical contrast between *discrete* (a line of points) and *continuous*.

Euclid operates with straight lines and circles. (a) His Parallel Postulate guarantees that two straight lines which are not parallel do *intersect*. He quietly assumes that a circle *intersects* (b) a straight line or (c) another circle, if either object is sufficiently near. As Euclid only works with those objects, this is all he ever needs in respect to continuity.

From the continuity of circles follows the continuity of ellipses, parabolas, and hyperbolas, for these figures are all *conic sections*.

Any cone emerges if a straight line rotates along a circle in space, one point fixed. Thus the continuity of the surface of a cone follows from the continuity of both the straight line and the circle.

A plane is continuous. Consequently, the line of intersection of a plane and a cone is continuous. But these lines of intersection are ellipses, parabolas, or hyperbolas, so the continuity of these lines follows from Euclid's assumptions. (These arguments are attributed to Menaechmus, a friend of Plato.)

Straight lines and circles can be drawn with the help of a ruler and a compass. The foregoing considerations show that conic sections can also be so constructed as continuous lines.

Synthesis: From Points to Curved Lines? (I)

As an outcome of his Analysis, Descartes found as the solution of the problem the equation

$$y = m - \frac{n}{z}x + \sqrt{m^2 + ox - \frac{p}{m}x^2}\,.$$

This equation shows how to find for each *given* indeterminate quantity x a root (or two roots?) y. These two (or three) numbers x and y denote lengths (*see* the figure on p. 15). As measuring starts from point B, x and y can also signify points on the "principal lines" AB and BC. We know how to work with coordinates: we draw a parallel to the principal line BC through the point x on AB; and likewise we draw a parallel to the principal line AB through the point y on BC. The result is *one* point. This point represents the pair of roots x, y, calculated beforehand.

But then—how do we get the line from the point?

Well, we do not have *one single* point, but *many:* to *each* number x we find a number y.

(Maybe not to *each,* as the expression under the square root must not be negative: but many; perhaps even infinitely many, as for every two admissible numbers— which should exist!—we have infinitely many others in between, and they, too, should be admissible.)

For that reason, the relevant question is: how do we extend the infinitely many points to become a line?

At this point Descartes becomes somewhat chary. He writes tersely:

> If then we should take successively an infinite number of different magnitudes x for the line [AB], we should also obtain an infinite number [of magnitudes] y for the line [BC], and, therefore, we have an infinity of different points, such as C, by means of which the required curved line could be drawn (descrire).

Or:

> Having explained the method of determining an infinite number of points lying on any curved line, I think I have furnished a way to describe them.

The Admissible Curved Lines

"… a way to describe them": *what* can Descartes "describe"? This he explains at the very beginning of his second chapter:

> Now to treat (tracer) all the curved lines which I mean to introduce here, only one additional assumption is necessary, namely two or more lines can be moved, one upon the other, determining by their intersection other curved lines.

With this characterization, and breaking with tradition, Descartes narrows the range of geometry (for him: mathematics): he only accepts *one single* movement for tracing lines.

> The spiral, the quadratix [...] must be conceived of as described by two separate movements whose relation does not admit of exact determination.

(A spiral is created when a radius rotates around a point while the pencil on it creeps outwards. The speeds of rotating and creeping are independent from each other, forming a completely arbitrary ratio. As Descartes is unable to perceive this ratio "clearly and distinctly", the spiral is not an admissible object in his mathematics. Analogously with the quadratix.)

The accepted lines are called by Descartes "geometrical" (for him the same as "mathematical").

Descartes gives another requirement for the admissibility of lines, namely

> that all points of those curved lines which we may call "geometric", that is, those which admit of precise and exact measurement must bear a definite relation to all points of a straight line, and that this relation must be expressed by means of a single equation.

In other words: Descartes only accepts those lines as "geometrical" ("mathematical") which can be described by only one equation.

Descartes' equations have the indeterminates only as summands, factors, or in the denominator of a fraction but never as exponents. Therefore, we can state, in today's language, that for Descartes the admissible ("geometrical") lines are the *algebraic* curved lines. Descartes excludes *transcendental* curved lines from his geometry (mathematics)—i.e. those whose equations have x in the exponent.

Synthesis: From Points to Curved Lines? (II)

Again: how do we get from the equation to the line? This is unclear and is still unclear even in Descartes' writings!

For, amazingly, Descartes was lacking in rigour regarding the transition from *already given points* to a continuous *line*. For example, he constructs the points of an oval with the help of the sections of two circles having F and G as centres (Fig. 2.2):

He finished his description of this procedure with the following statement:

> [...] we can thus find as many points as may be desired, by drawing lines parallel to 7 8 and describing circles with F and G as centres.

Today this may be sufficient (the reason for this will be understood later), but it was unacceptable in Descartes' time and did not meet classical standards. The intersection of circles around F and G—if they are near enough—is guaranteed by Euclid, and, therefore, the points of the oval which Descartes constructed do exist. But how do we know that one of these points is lying on the straight line AR? Maybe all of these intersections of the circles avoid this line? The figure does not seem to

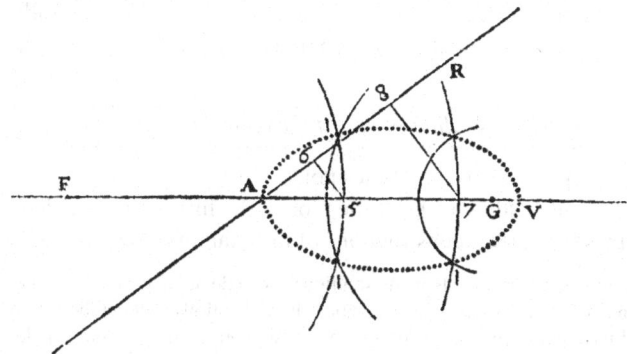

Fig. 2.2 Pointwise geometrical construction of an oval (*Discours* 1637, pp. 352, 358)

look like that—but this is not conclusive! Only if it were proved that the oval could be constructed would its intersection with the straight line AR be secured, as the meeting of two continua.

But Descartes is unable to draw this oval.

This shortcoming of Descartes is the more annoying for at the beginning of his book he accused classical mathematics of *exactly this* failure. Right at the beginning of his book he writes:

> Then, since there is always an infinite number of different points satisfying these requirements, it is also required to discover and trace the curved line containing all these points. Pappos says that when there are only three or four lines given, this line is one of the three conic sections, but he does not undertake to determine, describe, or explain the nature of the line required when the question involves a greater number of lines. He only adds that the ancients recognized one of them which they had shown to be useful [...] This led me try to find out whether, by my own method, I could go as far as they had gone.

"As far"—yes, but not further.

Descartes' Geometrical Successes and His Failure

In Chap. 1 we saw how Descartes embedded arithmetic (the numbers and the calculations with them) in geometry and that he invented the pure symbolic formula.

In this chapter we described the way in which Descartes *analyzed* a given problem of geometry, that is to say, how he translated it into his language of formulae, how he forms a suitable equation.

However, according to classical standards the *Analysis* of a problem—even if it is successful—is not sufficient. On the contrary, an Analysis has to be followed by a *Synthesis:* a construction of the problem from the components given by the Analysis.

But the Synthesis was Descartes' undoing, at least in general. He did not succeed in finding a method of tracing the line from the equation. We arrive at the surprising conclusion:

> Descartes did not invent the coordinate system.

Sometimes he succeeded, sometimes not.

In his very detailed study of Descartes' book the historian of mathematics Henk Bos summarizes Descartes' constructions of the examined curved lines as follows:

[...] the reconstructed arguments in the solutions above do not depend on the knowledge of the equations of the curved lines. They can be achieved entirely without the help of algebra. Moreover, I have not been able to devise a purely algebraic line of arguments leading in a natural way from the problems to the turning ruler and moving curve procedures for tracing the solution curves.

In short: Analysis done—Synthesis gone!

But *did* an alternative exist? Descartes' "*x*" are *numbers* or *line segments;* at best "*x*" and "*y*" allow the determination of *points*, maybe even infinitely many. But *points* are not *lines;* discrete objects are not continuous. Obviously, Descartes is lacking a *continuous x.*

Literature

Bos, H. J. M. (2001). *Redefining geometrical exactness. Descartes' transformation of the early modern concept of construction.* New York, Berlin, Heidelberg: Springer.

Cajori, F. (1929). Controversies on mathematics between Wallis, Hobbes and Barrow. *The Mathematics Teacher, 22,* 146–151.

Descartes, R. (1637). *Discours de la methode.* Leyde: Ian Maire. http://gallica.bnf.fr/ark:/12148/btv1b86069594.r=descartes+discours.lang.DE

Descartes, R. (1637). *Geometrie.* Darmstadt: Wissenschaftliche Buchgesellschaft, 1981. reprint of the 2nd edition 1923, Leipzig. German by Ludwig Schlesinger 1894.

Hobbes, T. (1662). Seven philosophical problems and two propositions of geometry. In: W. Molesworth (Ed.), *The English works of Thomas Hobbes* (Vol VII). London: John Bolton, 1845. second reprint: Aalen: Scientia Verlag, 1966.

Schmitz, M. (2010). *Analysis – eine Heuristik wissenschaftlicher Erkenntnis.* Freiburg im Breisgau: Karl Alber.

Stevin, S. (1586). *De Beghinselen der Weeghconst.* Leyden: Christoffel Plantijn. http://www.google.com/books?id=_wo8AAAAcAAJ&hl=de

Chapter 3
Lines and Variables

From Two to Infinity: Leibniz' Conception of the World

In Descartes we came to know a radical critic of contemporary rational thinking. With Gottfried Wilhelm Leibniz (1646–1716) we now turn to an equally radical diplomat who endeavoured to unite the world's antagonisms.

According to Leibniz, God created the "best possible world". His world consists of prime elements or atoms which Leibniz called "Monads". He says:

> The Monad, of which we shall speak here, is nothing but a *simple* substance, which enters into compounds. By 'simple' is meant 'without parts'.

All Monads differ from each other:

> Indeed, each Monad must be different from every other. For in nature there are never two beings which are perfectly alike and in which it is not possible to find an internal difference, or at least a difference founded upon an intrinsic quality.

Monads are beings, simple beings. Fundamentally,

> [...] nothing but this (namely, perceptions and their changes) can be found in a simple substance. It is also in this alone that all the *internal activities* of simple substances can consist.

Leibniz regarded it as a "metaphysical necessity"

> that every created being, and consequently the created Monad, is subject to change, and further that this change is continuous in each.

These central statements are to be found in the small booklet written in 1714, which was published four years after Leibniz' death as *Monadology*.

This way of thinking is radically antagonistic to Descartes. Whereas Descartes presumed two *substances* (extension and thought), Leibniz presumed (actual) infinitely many (the Monads). In other words, whereas Descartes had *principles*, Leibniz had *individuals*.

Leibniz' wider conceptions of the world are not connected to his mathematical thinking. The principal idea is: Monads are nothing else than *perception* as well as their *change,* an important assumption which underlies his entire philosophy. Even if he does not mention this explicitly, he intends it to *be* a fundamental fact. We will soon recognize this.

Leibniz' Mathematical Writings

Leibniz never worked as a regular mathematician. Nonetheless he is, together with Isaac Newton, one of the two creators of that mathematical theory which developed into the most powerful of all, till today: calculus.

Leibniz documented his invention in a lengthy essay; the longest mathematical text he ever compiled. It was supposed to be printed in Paris after Leibniz' departure from the town. But this did not happen and eventually the manuscript was lost. Leibniz, much later, refused to rewrite the paper: in his eyes, too much time had passed already.

However, among the vast amount of papers left by Leibniz, some drafts of this essay were found. An extract from such a draft was first published by Lucie Scholz in her dissertation in 1934, a complete version in 1993 by Eberhard Knobloch and meanwhile (2012) also in Leibniz' *Schriften*; a French translation by Marc Parmentier appeared in 2004 and a German translation by Otto Hamborg has been available since 2007 via the internet and since 2016 as a book.

Reading Leibniz' papers (which is not very easy as he wrote mostly in Latin) repudiates all the prejudices of present-day scientists against their predecessors: they were less intelligent than we are, their ideas were more vague, their reasonings inconclusive—only today are we qualified to be exact and precise.

The point is this: many earlier scientists did not use "vague" notions and "diffuse" reasonings—but *completely different* ones. If one engages with those *different* concepts and follows these *other ways* of reasoning, one realizes that Leibniz' proofs are not incorrect at all. Quite the opposite, Leibniz' ideas are marked by great ingenuity. His proofs, judged by today's standards, are as precise as his concepts allowed for.

Especially the following three concepts: convergence, integral and differential stand out as great scientific achievements.

Leibniz' Theorem: Fresh from the Creator!

The Convergence of Infinite Series

Up until today the basic curriculum of higher mathematics contains "Leibniz' Theorem". It is about infinite series, such as

$$1 - \tfrac{1}{2} + \tfrac{1}{4} - \tfrac{1}{8} + - \ldots$$

or

$$1 - \tfrac{1}{3} + \tfrac{1}{5} - \tfrac{1}{7} + \tfrac{1}{9} - + \ldots$$

As series are *infinite* sums, it is not possible to *calculate* them directly. Nevertheless, they often have a *value,* a *sum.* The two above happen to have a sum, but others do not. The series

$$1 - 2 + 3 - 4 + - \ldots$$

or

$$1 + 4 + 8 + 16 + \ldots$$

do *not* have a sum, that is to say: no *finite* value.

If one ponders for a while perhaps the following idea may arise: promising candidates for series with a sum may be those which fulfil two conditions: (i) the summands, their terms are ALWAYS *decreasing* and can become *as small as one likes them to be;* (ii) their *signs* alternate.

Consequently, the first term is greater than the sum, as the second term is *taken away* from it; the sum of the first two terms is smaller than the sum, as the third term is *added to it:* condition ii., etc. Finally: the *difference* (a *change!*) between the successively calculated sums decreases and becomes arbitrarily small: condition i.

This phenomenon that "the sum becomes increasingly *more accurate* if the differences between sums, which can truly be calculated, decrease below each given quantity" is what we call "convergence" today. Yet, during Leibniz' time this name was not established.

> Leibniz was the first ever to describe these novel objects of mathematics ("convergent series") systematically.

Leibniz' Formulation of His Theorem

In his manuscript Leibniz wrote:

If a quantity A is equal to a series $b - c + d - e + f - g$, etc.,

$$A = b - c + d - e + - \ldots,$$

which decreases infinitely in such a way that the terms eventually become smaller than an arbitrarily given quantity, it will be

$+b$	greater than A, so that the difference is smaller than	c
$+b - c$	smaller ...	... is smaller than d
$+b - c + d$	greater ...	... smaller than e
$+b - c + d - e$	smaller ...	... smaller than f.

And in general, the part of the decreasing series with alternating additions and subtractions, which ends with an addition, will be greater than the sum of the series, the part which ends with a subtraction will be smaller; but the error or the difference will always be smaller than the term of the series, which follows the part at once.

With the exception of the line "$A = b - c + d - e + - \ldots$" and the last three "is smaller than", this is *exactly* what Leibniz wrote in his manuscript. It is as precise as possible. Leibniz describes *in all detail* what an "infinite series" with "alternating signs" and "steadily decreasing" terms which "decrease below each assumed quantity" is.

It is permissible *to read* this statement of Leibniz as follows:

Theorem. *If the terms of a series*

$$b - c + d - e + - \ldots$$

decrease indefinitely (i.e. they eventually get less than any assumed quantity) than this series has a finite sum.

Leibniz did not leave it at this description but added a fairly detailed proof.

Leibniz' Proof of His Theorem

Leibniz' proof consists of a preliminary consideration and four steps.

Preliminary: as the amount of the terms steadily decreases, there is *altogether* more added than subtracted. Added were $b + d + f + \ldots$, subtracted were $c + e + g + \ldots$ and we have $b > c, d > e, f > g$, etc. In this manuscript Leibniz does not use a "greater" sign, but in others he does.

Especially we have for the sum A:

$$A < b.$$

1. The first step of the proof: using the last inequality we may consider the number

$$b - A.$$

(We see: Leibniz tries to deal with positive, *true* numbers!) Therefore,

$$b - A = b - (b - c + d - e + f - g + - \ldots) = c - d + e - f + g - + \ldots < c,$$

where the last inequality $<$ holds for the same reason as was given for $A < b$ in the preliminary!

2. The second step of the proof: we found that $A < b$. *In the same way* we derive from (1) that

$$A > b - c.$$

Is it?—Really!—Therefore, it is allowed to take $A - (b - c)$ (which is also a true number) and we get

$$A - (b - c) = d - e + f - g + - \ldots < d,$$

by the same argument.

3. Leibniz' third step of the proof: as before, through rearranging the result above, we get as the next starting point:

$$A < b - c + d.$$

So we can build $(b - c + d) - A$. And by the same procedure we arrive at

$$b - c + d - A = e - f + g - + \ldots < e.$$

Leibniz adds a fourth step in the same manner, but we need not repeat it here, as the pattern is clear by now.

As a result Leibniz gets a sequence of inequalities:

$$A < b$$
$$b - A < c$$
$$A - (b - c) < d$$
$$(b - c + d) - A < e$$
$$A - (b - c + d - e) < f$$

$$\ldots$$

From the second line onwards we have on the left side the *error* that arises, if instead of the whole series only its beginning till the nth term is taken and the right hand side shows that this error is *always* less than the $(n + 1)^{\text{st}}$ term. But it was presupposed that the terms' magnitude "eventually decrease below any assumed magnitude", and, therefore, this assumption—as can be seen in the above inequalities—leads to the conclusion that the *the error caused by breaking off* the series also becomes less than any assumed magnitude.

What more do we want?

> It is impossible to give a more precise, *stronger* argument.

And that even today. Leibniz did so in his manuscript of late summer 1676.

Reflection on Leibniz' Achievement

Through these notes Leibniz became the first person in the history of mathematics to describe, *as precisely as possible,* what is nowadays called the "convergence" of an "(infinite) series". (The attribute "infinite" is plainly superfluous but it sounds so breathtakingly bombastic.)

The expression "convergence" was not used by Leibniz. However, the importance is the content of the finding, the name is quite irrelevant. What counts is that Leibniz stated his result in absolute clarity.

The philosophical question: what did Leibniz do? What was the object of his inquiry?

My answer: obviously, Leibniz explored a *variable quantity,* i.e. the *successively calculated sum* of a series. Today this is called the "partial sum" and we have a standard symbol for it: s_n. We write:

$$s_1 = a$$
$$s_2 = a - b$$
$$s_3 = a - b + c$$
$$s_4 = a - b + c - d$$
$$s_5 = a - b + c - d + e$$
$$\dots$$
$$s_n = a - b + c - + \dots \pm n$$

It is evident that this object, the "partial sum" or s_n, is a *variable* (i.e. changing) quantity. Leibniz only had the right hand sides of the above equations and, therefore, they can be named without any reservations ("s_n" will do as well as any other name).

Without explicitly saying so, Leibniz studied *variable* quantities.

Leibniz' way of arguing is so fascinating, so deeply mathematical that nobody dared to characterize it as "unmathematical" and to criticize it: According to the standards of classical mathematics this is not mathematics!—At least to our knowledge nobody dared to make such an accusation, written or otherwise.

An Idea Which Leibniz Could not Grasp and the Reason for His Inability

Today we illustrate the above facts by the following picture:

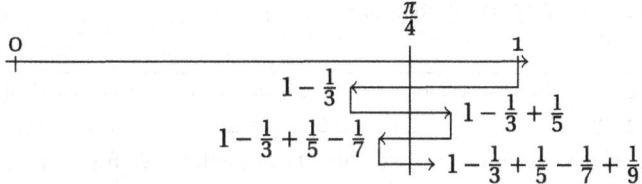

At a single glance we *grasp* the situation. However,

Leibniz was unable to draw this picture.

For this picture demands two things: (a) lengths have *a direction,* and (b) "negative" numbers *do exist.*

Actually we have seen earlier that a length has no direction! Moreover, if negative numbers were true "numbers", some of the more than two thousand years old laws (known at least since Euclid!) would be invalidated.

One of these laws reads as follows:

$$\text{If} \quad \frac{a}{b} = \frac{c}{d}, \quad \text{and if} \quad a > c, \quad \text{then} \quad b > d \, .$$

But if -1 is a "number", this law requires:

$$\text{As} \quad \frac{1}{-1} = \frac{-1}{1}, \quad \text{and if} \quad 1 > -1 \quad \text{it follows} \quad -1 > 1 \, .$$

and thus a contradiction. Contradictions are *absolutely* forbidden in mathematics for otherwise *everything* whether false or correct can be proved.

The mathematicians of the late seventeenth century and the beginning of the eighteenth century had to make a *decision:* should those time-honoured laws be preserved, or should −1 become a true *number?*

Leibniz was an astoundingly creative thinker but not a revolutionary; he was a diplomat. He shunned revolutions (the dismissal of the validity of these classical laws) but praised the news by couching it in unctuous words:

> Nevertheless, I do not want to deny [...] that −1 is a quantity smaller than nothing; this only has to be understood right-minded. Such statements are what I call *passable true* (following the renowned Joachim Jungius); [...] However, they would not bear a severe verification, but yet they are of great help for the calculation and of immense value for the inventive genius as well as for universal concepts.

The classical phraseology of diplomats: "... / as well as ..."; "−1 is smaller than 1 / but this has to be correctly understood"; "to be on the safe side, I cite an authority, however, unknown or vague, or hint at my obedience: strictly speaking this is forbidden / but it is of huge utility".

Thus, the paper published by Leibniz in a famous scientific journal in April 1712 can be understood as an act of great political diplomacy.

The quintessence is:

> Accepting "negative" numbers as *true* ones goes hand in hand with the invalidity of some time-honoured laws.

If he or she may say something like: "this law is valid *only* for positive numbers, and all is fine", the known contradictions are outlawed. (Hopefully, no others will appear we have not yet thought of!)

The Precise Calculation of Areas Bounded by Curves: The Integral

The Beginning Is Easy

Lasting for thousands of years, the Babylonians operated quite differently compared with classical Greek scholars, but in Greek culture the following art of planimetry was taught:

It only allowed for *rectangles.* The area is

length times width.

width

length

From this all further calculations had to be deduced, e.g.: the area of a *rectangular* triangle is

$$\frac{1}{2} \text{ times base times height.}$$

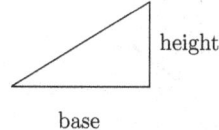

height

base

The Problem

What happens, if the boundary of one side is curved?

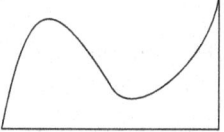

The Solution of Leibniz—The Original Way

For such an instance Leibniz has the following astounding idea:

I present Leibniz' original figure, omitting what is not essential to the principal idea. Even so, the figure is still sufficiently complicate (Fig. 3.1).

The object of the construction is the area between the curve D_1, D_2, D_3, D_4 and the three line segments $\overline{D_1 B_1}$, $\overline{B_1 B_4}$, and $\overline{B_4 D_4}$. The points D_n are *any* points on the curve.

1. Leibniz takes the stairway B_1, N_1, P_1, N_2, P_2, N_3, P_3, B_4, B_1 as a *first approximation* of the area. In the following and for the sake of simplicity we will indicate the single steps of the stairway by their dotted "upper" lines, the "step zones", i.e. $N_1 P_1$, etc.

2. Now let us assess the *error* of this approximation! It consists of the sum of *partial errors*. Compared with the actual area we get:

   (a) The first step $N_1 P_1$ is *too big* by the (curved) triangle $D_1 N_1 F_1$—as well as *too small* by the triangle $F_1 P_1 D_2$. Leibniz is saying that at any rate this *first* partial error is *less than the whole rectangle* in between D_1 and D_2. This is *really* generous, isn't it!

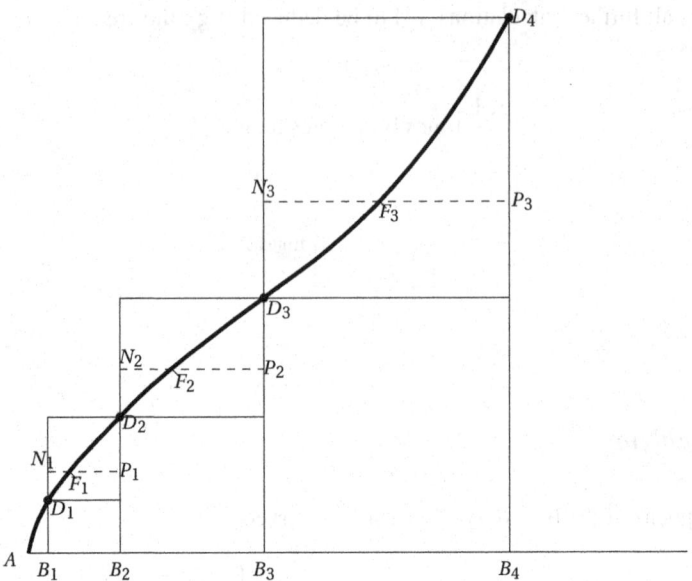

Fig. 3.1 Leibniz' figure to calculate the area (purified), 1676

(b) The same with the following step $N_2 P_2$, again the *second* partial error caused by the approximation is clearly *less than the rectangle* in between D_2 and D_3.

(c) And so forth.

3. So what is the *total error* at most? Most likely it is less than the *sum* of the partial errors; undoubtedly it is less than the sum of the rectangles D_1, D_2; D_2, D_3 and D_3, D_4.

4. Therefore, what is our estimate of the *total error?* The total *height* is obviously (if the curve is always ascending or always descending; otherwise it must be divided in such way) the height from D_1 to D_4. And, to be on the safe side, Leibniz chooses as *width* the *maximal* width of the steps $\overline{B_1 B_2}$, $\overline{B_2 B_3}$ and $\overline{B_3 B_4}$. In the example it is $\overline{B_3 B_4}$.

Outcome: the *total error* of the first approximation is clearly less than the *product of the height from D_1 to D_4 and the maximal width of the* $\overline{B_n B_{n+1}}$.

5. All boils down to the following conclusion:

The points D on the curve are completely arbitrary. We may choose as many of them as we please. Let's say, we choose k equidistant points such that the steps have width $\overline{B_1 B_k}/(k-1)$.

Consequently, the *total error* of this k^{th} approximation will be certainly less than the rectangle with width $\overline{B_1 B_k}/(k-1)$ and height $\overline{D_1 D_k}$ (more precisely, $\overline{B_1 D_k}$).

Whereas the height of the rectangle remains the same, its width *continuously decreases* if further points D are chosen. And the *total error* of the kth estimate is clearly less than the area of the corresponding rectangle.

Subsequently, the product of these two values will also drop: if in the product $b_k \cdot h$ the factor b_k is constantly decreasing while h (the height) remains the same, the product decreases, too.

Thus the total error of the estimate reduces further.

6. How small does it become?

Obviously, there is no lower limit to the area's magnitude (actually, its "smallness"): with the exception of the limit zero, of course. According to Leibniz' own words:

> The points D may be thought of as near and in such a great number that the straight-lined step-shaped area differs from the four-lined area $D_1 \, B_1 \, B_4 \, D_4 \, D_3$ etc. D_1 itself *by a quantity which is less than an arbitrarily given.*

This means: the *total error* which emerges from the calculation of the area below the steps instead of the area limited by the curve, can be *decreased* to any desired degree of exactitude.

Did we hit the jackpot? Do we have the area?—Yes and no. On the one hand, we have a *method of calculation:* Leibniz is capable of calculating the area as precisely as he wants to.

On the other hand, being able to calculate something does not mean to have a concept for doing it. Engineers may be satisfied with a method of calculation, but mathematics needs concepts! In this case, an appropriate concept of numbers. However, Leibniz could not offer one. Understandably so, as it turned out mathematics needed two further centuries to coin such concepts (Chaps. 13 and 12) and in a certain sense, these solutions contradict Leibniz' thinking, for they demand the acceptance of the "actual infinite" in mathematics (*see* Chap. 5).

Outlook

Leibniz' idea from 1676 provided the foundations for what was later called "integration", especially for the "integral" of a "function", the graph of which Leibniz still called "curved line".

However, the names "integral" as well as "function" were already used by Leibniz, but both with other meanings.

(a) "Function" was Leibniz' title for a multitude of line segments which can be constructed to make up a curve when set in relation to straight coordinate axes: abscissa, ordinate, tangent, normal, sub-tangent, sub-normal, resecta, ..., a lot of special geometrical constructions, which were frequently studied in his times.

> Leibniz used the name "function" without a definite meaning.

(b) The name "integral" was invented by Johann Bernoulli and in 1690 his brother Jacob used the notation for the first time in a printed paper. The first printed document in which Leibniz used the *sign* "$\int$" dates back to the year 1686. He used it in connection with his sign for the "differential": "dx"—a concept which will be treated in the next section. As "d" is an operator, "dx" as well as "dy" are to be read as a *single* quantity in the following text.

> If one transforms the differential equation $p\,dy = x\,dx$ into the "summatorial" [by building the sums on both sides], one has $\int p\,dy = \int x\,dx$. From what I have shown in the method of tangents, we clearly have $d(\frac{1}{2}xx) = x\,dx$; therefore, the reverse is $\frac{1}{2}xx = \int x\,dx$ (because like powers and roots in usual calculations, we have sums and differences or $\int$ and d as reciprocal).

The first evidence of the integral sign, as it is still used today, "$\int$", goes back to 1691. Leibniz had encountered the name "integral" for the first time in 1690, in an article by Jacob Bernoulli

However, the *topic* that was presented above, i.e. the exact calculation of an area with a curved boundary, was made the subject of mathematics, *as precisely as 1676,* only 178 years later. According to its later inventor Bernhard Riemann, the mathematical object is nowadays known as the "Riemann integral". Shortly before Riemann, Cauchy had come up with a similar idea (pp. 145f).

To sum up: Leibniz had already developed this notion *as precisely as possible*— but without today's concepts of "function", "infinite series", and "convergence". It worked without these notions!

If Leibniz' plan had succeeded and his manuscript had been printed, mathematics would have developed differently.

Leibniz' Neat Construction of the Concept of a Differential

The First Publication: A False Start

Leibniz published his idea of "differential" from the 1670s in October 1684. Albeit, his *explanation* remained very vague. Even worse, Leibniz made a mistake such that the intelligent reader had to decide whether the whole treatise was wrong or merely the definition of its fundamental notion. In case of the latter, the reader had to reconsider the *correct* concept of "differential" all by himself.

Clearly, this publication was almost a complete failure. Where to find a clever and astute reader? Jacob Bernoulli (1654–1705) had tried hard to understand this text since 1687. In the end he needed two or three years to understand the main idea in order to use the new method himself. What's more, he was able to develop it further, in dialogue with his younger brother Johann—and with Leibniz.

Another False Start: The New Edition

The differential calculus consists of the *concept* of "differential" as well as *laws of calculation* for those differentials. Without these laws the concept is of no use.

Leibniz owed the explicit formulation and *the detailed foundation* of these laws to the public of his time. Just as before, he wrote a manuscript thereon but did not publish it. In 1846, when it was finally published, together with a lot of other manuscripts on this topic, nobody took notice. The same happened in 1920, when the English translation appeared. This fact has been documented only fairly recently: by Henk Bos in 1972 and by Richard T. W. Arthur in 2013, who are both historians of mathematics.

The Neat Construction, Part I

In regard to the above presentation it comes as no surprise that Leibniz took also the "differential" to be a *geometrical* notion. More precisely, with help of the "differential" it should become possible to draw a *tangent* to an arbitrary curved line (Fig. 3.2).

A "tangent" is a straight line that *touches* the curve, i.e. that snuggles up to it. *To touch* usually means there is *no cut,* the curve remains on the one side of the straight line. However, there are unusual curves and *sometimes* the "tangent" still cuts those curves.

Leibniz came up with the following geometrical construction of the differential. The abscissa x is drawn upwards, the ordinate y to the right (today we usually do it the other way round). Take a parabola

$$y = \frac{x^2}{a} \, ,$$

Fig. 3.2 Leibniz' calculation of the tangent

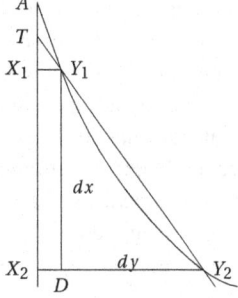

which is represented by the curved line in the figure. Leibniz chooses $AX_1 = x$ and $X_1Y_1 = y$. From the point Y_1 the perpendicular line Y_1D to the larger horizontal line X_2Y_2 (the ordinate) is drawn. The *difference* of AX_2 and AX_1 is called by Leibniz the "differential" dx, similarly, the *difference* of X_2Y_2 and X_1Y_1 is called the "differential" dy. These are the notations, now the calculations:
The equation of the curve reads

$$y = \frac{x^2}{a} .$$

Leibniz starts by *changing* x to $x + dx$ and subsequently y to $y + dy$. This modifies the equation to:

$$y + dy = \frac{(x + dx)^2}{a} = \frac{x^2 + 2x\,dx + dx^2}{a} .$$

We subtract the original equation and get:

$$dy = \frac{2x\,dx + dx^2}{a} = \frac{2x + dx}{a} \cdot dx \qquad \text{or} \qquad \frac{dy}{dx} = \frac{2x + dx}{a} .$$

Of course, dx and dy are *changing* quantities. And as we would expect they *decrease indefinitely: below any given quantity.*

Therefore, the numerator of the fraction will *approach* $2x$. But then, as zero is not allowed as denominator, we encounter a problem regarding the dx on the left side of the equation. Yet, if dx does not *truly* reach its limit zero, the numerator on the right will not become $= 2x$.

What are we to do?

Interlude: The General Rule: The Law of Continuity

Leibniz needs an argument that enables him to *extend* the validity for cases only holding for $dx \neq 0$ to the case $dx = 0$.

And indeed, Leibniz really had such an argument at his disposal: his Law of Continuity. According to the historian of mathematics Herbert Breger, this cognitive law has a similar significance and power for Leibniz as later the Method of Dialectics has for Hegel. It is a "universal scheme of thought and cognition". And as we might expect, coming from Leibniz, its main characteristic is a diplomatic rather than a logical one. The Law of Continuity *unites opposites* instead of separating them as logic does.

> Generally speaking, Leibniz' Law of Continuity says: different things, even if they are contradictory, can emerge (or be understood) from a common principle.

This is the principal idea. Naturally, the law may be given some more precise formulation, in accordance with the concrete requirement.

Thus, in order to give his concept of "differential" a rigorous foundation, Leibniz formulated his Law of Continuity in his manuscript as follows:

> If some continuous transition ends in some limit, it should be allowed to state a common law of thought that includes the last limit.

This bails him out.

The Neat Construction, Part II

Leibniz takes the differential triangle $Y_1 D Y_2$. It is his idea to oppose it to an auxiliary triangle with two features: (a) It is *similar* to the differential triangle. (b) One of its sides is *fixed*.

This auxiliary triangle originates by producing the side $Y_2 Y_1$ to T on the axis $A X_2$.

The auxiliary triangle $T X_1 Y_1$ has the same angles as the differential triangle and is, thus, similar to it.

Then point Y_2 moves on the curved line (in our example: the parabola) toward the point Y_1. What is going to happen?

Point T moves along the vertical axis $A X_2$ up and down, whereas the side $X_1 Y_1$ remains fixed. However, the differential triangle $Y_1 D Y_2$ and the auxiliary triangle $T X_1 Y_1$ remain similar (if the latter triangle degenerates, one has to think anew).

Next, if point Y_2 coincides with point Y_1, we have exactly the situation which is covered by the Law of Continuity in its above specification: the coinciding of the points Y_2 and Y_1 represents the *limit;* the case in which the differential triangle has *vanished.* However, the auxiliary triangle survives the coinciding of Y_2 and Y_1—for its side $X_1 Y_1$ is *fixed* and cannot vanish. Therefore, *this auxiliary triangle $T X_1 Y_1$ represents the* COMMON PRINCIPLE *which covers* BOTH CASES *the vanishing differential triangle* AND *its disappearance.*

Both triangles are similar. Consequently, we have

$$\frac{D Y_2}{Y_1 D} = \frac{dy}{dx} = \frac{X_1 Y_1}{T X_1}.$$

Leibniz already knows: $\frac{dy}{dx} = \frac{2x+dx}{a}$. So all together he has:

$$\frac{dy}{dx} = \frac{X_1 Y_1}{T X_1} = \frac{2x + dx}{a}.$$

Actually, Leibniz can apply his Law of Continuity: on the left there is a fraction in which numerator and denominator vanish together; the middle is a fraction with changing value but both, numerator and denominator, *remain* and stay *finite;* on the right there is a fraction with *vanishing* dx in the numerator but all other quantities remain fixed.

Next, Leibniz lets the dx vanish, i.e. $dx \to 0$. The outcome is obvious:

$$\frac{dy}{dx} = \frac{2x}{a} \,,$$

which is with the help of his Law of Continuity absolutely neatly derived—although on the left is a fraction where numerator as well as denominator tend toward zero!

This is exactly the point! Leibniz does not divide 0 by 0 but analyzes a ratio $\frac{dx}{dy}$ with *simultaneously* (or, at the same time—but we should keep time out of mathematics!) *vanishing* terms. In today's notation: $\lim\limits_{dx \to 0} \frac{dy}{dx} = \frac{2x}{a}$. The auxiliary triangle provides Leibniz with a fixed point to unhinge the world. The Law of Continuity, invented by him, allows him the precise calculation of this ratio dy/dx even in the case where $dx \to 0$ as well as $dy \to 0$.

What Is x (and What Is dx) for Leibniz?

Leibniz develops the differential as a geometrical concept. He denotes the respective *length* on the x-axes by "x". Of course, this length is not invariable—just the opposite! For Leibniz the length x is *changing* and this change is *described* by "$x + dx$".

> For Leibniz *both x and dx are varying lengths.*

We know, Descartes denoted by "x" a certain fixed number or length. Leibniz revolutionized Descartes' world of concepts completely! Only the letter "x" survived—as if nothing had happened. But an upheaval took place: the *continuum* was introduced in the calculations on the quiet.

Of equal importance is that Leibniz created dx as *a changing quantity which decreases below any given quantity.* This circumstance was named by him an "infinitely small quantity".

Consequently, an "infinitely small quantity" is for Leibniz nothing supernatural, inconceivable—but only a *special case* of a commonly used changing quantity: just one which *decreases indefinitely* (although Leibniz had no "negative" numbers).

Using the concept "limit" one may say: for Leibniz an *"infinitely small quantity" is a changing quantity with zero as its "limit".*

Literature

Arthur, R. T. W. (2013). Leibniz's syncategorematic infinitesimals. *Archive for History of Exact Sciences, 67*, 553–593.

Bernoulli, J. (1690). *Op. XXXIX. Analysis problematis antehac propositi*. In Opera, S. T. (Vol. 1), pp. 421–426. Genevæ: Hæredum Cramer & Fratrum Philbert, 1744. Acta Erud., Mai 1690, pp. 217–219; cf. Bos 1974, p 21.

Bos, H. J. M. (1974). Differentials, higher-order differentials and the derivative in the Leibnizian Calculus. *Archive for History of Exact Sciences, 14*, 1–90.

Breger, H. (2016). *Kontinuum, Analysis, Informales – Beiträge zur Mathematik und Philosophie von Leibniz*. In W. Li (Ed.). Berlin: Springer Spektrum.

Buchenau, A. (1966). *Gottfried Wilhelm Leibniz – Hauptschriften zur Philosophie* (Vol. 2). In E. Cassirer. Vols. 107, 108 of the series Philosophische Bibliothek. Hamburg: Felix Meiner, 1904–1906.

Gerhardt, C. I. (Ed.) (1846). *Historia et Origo Calculi Differentialis a G. G. Leibnitio conscripta*. Hannover: Hahn'sche Hofbuchhandlung. https://archive.org.

Hofmann, J. E. (1966). Leibniz als Mathematiker. In W. Totok & C. Haase (Ed.), *Leibniz, sein Leben – sein Wirken – seine Welt*. Hannover: Verlag für Literatur und Zeitgeschehen.

Knobloch, E. (Ed.) (2016). *Gottfried Wilhelm Leibniz: De quadratura arithmetica circuli ellipseos et hyperbolae*. Klassische Texte der Wissenschaft. Berlin: Springer. German by Otto Hamborg.

Latta, R. (1898). *Leibniz. The Monadology and Other Philosophical Writings*. Oxford: Clarendon Press.

Leibniz, G. W. (1982). Monadologie. In *Gottfried Wilhelm Leibniz: Vernunftprinzipien der Natur und der Gnade. Monadologie*. Vol. 253 of the series Philosophische Bibliothek (pp. 26–69). Hamburg: Felix Meiner. German by Herbert Herring 1956. Also in: Buchenau 1904–06, Vol. II, pp. 435–456, 1714

Leibniz, G. W. (1993). De quadratura arithmetica circuli ellipseos et hyperbolae cujus corollarium est trigonometria sine tabulis. In E. Knobloch (Ed.), *Abhandlungen der Akademie der Wissenschaften in Göttingen*. Göttingen: Vandenhoeck & Ruprecht.

Leibniz, G. W. (2004) *Gottfried Wilhelm Leibniz: Quadrature arithmétique du cercle, de l'ellipse et de l'hyperbole et la trigonométrie sans tables trigonométriques qui en est le corollaire*. Lat. text: Eberhard Knobloch (Libraire Philosophique. Paris: J. Vrin), ed.: Marc Parmentier.

Leibniz, G. W. (2011). *Die mathematischen Zeitschriftenartikel*. In H.-J. Heß & M.-L. Babin (Eds.). Hildesheim, Zürich, New York: Georg Olms Verlag.

Leibniz, G. W. (2012). De quadratura arithmetica circuli ellipseos et hyperbolae. In Leibniz-Archiv Hannover (ed.), *Gottfried Wilhelm Leibniz – Sämtliche Schriften und Briefe*. Vol. 6 – 1673–1676, Arithmetische Kreisquadratur, of the series VII Mathematische Schriften, pp. 520–676. N 51 (1676). https://www.gwlb.de/leibniz/digitale-ressourcen/repositorium-desleibniz-archivs/laa-bd-vii-6.

Scholtz, L. (1934). *Die exakte Grundlegung der Infinitesimalrechnung bei Leibniz*. Dissertation thesis, Hohe Philosophische Fakultät der Philipps-Universität Marburg. Görlitz-Biesnitz: Verlagsanstalt Hans Kretschmer.

Chapter 4
Indivisible: An Old Notion (Or, What Is the Continuum Made of?)

A Modern Theory?

Gottfried Wilhelm Leibniz invented a theory and the language of this theory is still used in modern mathematics. Today, just like him, we write differentials as "dx", "dy" and integrals as "$\int y\ dx$".

However, today these *signifiers* do not have the same meaning as they had for their daring inventor. (Let's also remember that the sign "$\int$" was not at all invented by Leibniz but by his younger colleague Johann Bernoulli.) Whereas Leibniz thought of areas and lines as purely geometrical quantities, these are seen today as much more general concepts. Nowadays, the differential "dx" is often only a symbol of calculation without any material significance—whereas Leibniz thought it, as we have seen, to be a "variable (geometrical) quantity", "decreasing indefinitely", "below any given quantity", and "eventually vanishing".

The transition from Descartes to Leibniz showed us a complete semantic change of the "x" in the formulae! Descartes thought the "x" to be an unknown "number" which was to be calculated—but essentially for him it signified a definite length of a line segment as he held arithmetic at bottom to be the same as geometry (although in some new version, e.g. a "unit" had to be defined). To sum up, Descartes thought of the "x" as a "number" or as a "line segment"—whereas Leibniz made "x" to be a "variable quantity", or in short: a "variable".

In the following chapter we shall learn in which way Leibniz' concept of differential was developed and changed by Johann Bernoulli. We shall come to this later and note for now:

> The foundational concepts of mathematics, with which we are dealing, change over time.

So one has to be careful not to superimpose our modern understanding on an old text. At least, we should try not to, as it is all too easy to fall into ingrained habits. Let's see!

© The Author(s), under exclusive license to Springer Nature Switzerland AG 2022
D. D. Spalt, *A Brief History of Analysis*,
https://doi.org/10.1007/978-3-031-00650-0_4

Leibniz Knew His Theory Was Descended from an Old Tradition

Leibniz formulated the foundations of his new theory during the years 1674–76 in Paris. Naturally, he continued working on them. Because of his first publication on the topic in 1684, which was hard to understand (we mentioned this on p. 32), two very capable mathematicians had become alert: the brothers Jacob and Johann Bernoulli. Subsequently, this triumvirate developed Leibniz' fundamentals to a great extent.

As yet in the year 1692, Leibniz called his system "our *analysis of indivisibles*". Thus, nearly twenty years after his invention he stuck to his original methodology: the "method of indivisibles". What this is about will be indicated in this chapter. Only indicated, for the details are too intricate to be explained here. But the basic ideas are of great importance: we need to realize that Leibniz did not create his theory from scratch. The studies of former scholars opened up the path toward Leibniz' creation. Nevertheless it was he who developed these studies further and into another direction which turned out to be so very successful. In a similar way, the same holds for Newton.

Before we turn toward the concept of "indivisible", we have to focus on an issue which is even older, and which has the name "continuum". Today we still work with the notion of "continuum" but do not use "indivisible".

The Continuum and Why It Does Not Consist of Points

What Is the Continuum?

The continuum is cohesive, unbroken, connected. Prototypes are the line, the area, the volume as well as the course of time—Thus far, very well and easy.

But notice, although the continuum is cohesive, it can be *divided*.

The course of time is divided by the present into past and future. The line is divided by the point in "left of" and "right of" it. The area is divided by a whole line. This, too, is self-evident and obvious.

For the future we register:

> The continuum can be divided.

And it is also evident, what the continuum divides is *inside* the continuum. The present is a moment in time. The point, the area are limits within the continuum. In short, what *divides* the continuum *belongs* to it.

How Do Continuum and Point Interact?

Now the question: *What is the proper relation between point and continuum?*

Clearly, *the point belongs to the continuum.* The moment belongs to the course of time, etc. But the other way round? Do *only* points make the continuum?

Maybe you will answer this question with "Obviously!". What else should exist within the continuum?

But it is not that simple! Caution: point and continuum differ from each other in an *essential* aspect: the continuum can be divided, as we have just pondered upon— but the point cannot be divided.

> A "point" is that which has no part.

This is the definition of "point". It is already in Euclid. For him it is the first definition, the very first sentence at all.

But what has no parts, clearly cannot be divided.

Therefore, continuum and point are essentially different: the continuum can be divided; the point cannot.

What follows from this?

The Continuum Does Not Consist of Points

The pair of concepts "part/whole" was already the subject of the earliest and most elementary aspects of philosophical thought in occidental culture. Early philosophers used these notions to speculate about the metaphysical nature of the world. One of the bedrocks of these ideas is the following truism:

> "Part" can only be what has the *essential qualities* of the whole.

This principle remained unchallenged in Western philosophy up to the beginning of the twentieth century, or until the development of set-theory.

As long as we accept this foundational axiom of philosophical thought (at least for Western culture), we arrive at the result above and have to *conclude* the following theorem:

> The continuum does not consist of points (or "nows").

Let us recapitulate the proof! It is made up of four steps:

Fact 1.   The continuum can be divided.
Fact 2.   The point cannot be divided.

Fact 3.   The continuum has a quality which the point does not have: divisibility. This quality is an *essential* one because for the continuum it is *essential* that it can be divided. Everything which is extended in space or persisting in time *can* be divided.

Fact 4.   *Consequently,* the point *cannot* be "part" of the continuum—End of proof!

We have shown *conclusively* that: "The continuum does not consist of points (or nows)". It was not terribly difficult to prove our initial statement! Or was it?

For thousands of years this way of thought has been accepted within occidental culture until, some 150 years ago, it became outdated. It was held no longer suitable for the new times and thrown onto the rubbish heap. We shall return to this later, starting from Chap. 12. However, one fact can already be mentioned here: the later rise of set-theory quickly did away with this age-old way of thinking.

But we are not done yet! There are still some further developments of analysis and, most importantly, a short retrospective regarding the late Middle Ages, to which we will turn now.

The Indivisible

We remember: Leibniz, as late as 1692, spoke of his theory as "our *analysis of indivisibles*". This Latin-based name was taken over into the English language; in German it sounds a little arcane today. As long as Latin remained the language of scholars (and theologians), such notions were widely accepted.

Thomas Aquinas

Thomas Aquinas (*c*1225–74) was a philosopher and theologian and one of the most famous and influential scholastics of the late Middle Ages. Aquinas used the notion "indivisible" when he spoke of a point or an instant of the spatial or temporal continuum: the indivisible is the point on the line or the present moment in time.

Nicholas of Cusa

There were other philosophers for whom the concept of the indivisible played an important role. One of the most illustrious is Nicholas of Cusa (1401–64) whose Latin name is Nicolaus Cusanus.

Although Leibniz lived about 250 years later than Cusanus, there are many similarities between the two men. Both were scholars of jurisprudence as well as diplomats and both were keen travellers. Nicholas of Cusa was a modern scholar of his times. Just as Leibniz much later, Cusanus was a pioneer of new thought. One of his revolutionary mottoes was: "Man is the measure of all things!" In reawakening this doctrine (usually attributed to Protagoras) he opposed the pious tradition of his contemporaries.

He formulated a principle of thought which is called "Doctrine of Coincidence". Somewhat shortened, it says: reason is the wholeness of those opposites (including contradictions!), which are incompatible to our understanding. At once we are reminded of Leibniz' Law of Continuity (p. 34).

Those mathematicians who think that a wholeness of contradictions is an evil trick may be referred to the quotation of a contemporary philosopher (who is a great authority on Cusanus), Kurt Flasch:

> Who once got acquainted with the contradiction that our thinking *is,* will grasp that the Law of Noncontradiction cannot be a philosophical criterion of truthfulness.

Because, says Flasch:

> Thinking is rest as well as motion; both are its qualities; who tries to make distinctions between them, in order to get rid of the contradiction, damages the elementariness of thinking.

(Caution: This "elementariness" does not mean "simplicity" but is to be understood as the Leibnizian designation: "elementariness" means "without parts"— p. 21.)

Let me present at least one sentence from Nicholas of Cusa on the indivisibles, to be found in his *Conjectures* written *ca.* 1440/44:

> However, reason is of such a lucid nature that it grasps, so to speak, the whole sphere in its indivisible center.

The "indivisible centre" is the "puncto centrali indivisibili".

Buonaventura Cavalieri

The (younger) contemporary of Galilei, Buonaventura Cavalieri (1598?–1647), made the "indivisible" a principal notion of his mathematical theory of the calculation of areas. This theory is explained in two books. Cavalieri came too early and could not make use of the language of formulae which Descartes was to publish in 1637, and consequently, his writings are not easily understood by us.

However, the historian of mathematics Kirsti Andersen took on the challenge of thoroughly deciphering Cavalieri's writings. Her representation will be my source in what follows.

1. In a letter to Galilei from 2nd October 1634, Cavalieri wrote clearly: "I absolutely do not declare to compose the continuum from indivisibles".
2. It is Cavalieri's principal idea to *compare* areas and to draw conclusions from these comparisons regarding the magnitudes of those areas.

Fig. 4.1 Cavalieri: straight
and oblique traverse
(*Exercitationes* 1647, p. 15)

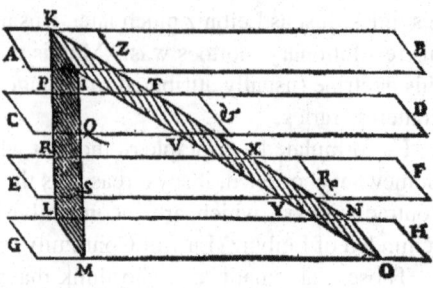

He traverses a "ruler" through two areas and *compares* what happens. While
doing so he is interested in a concept invented by him and called "all the lines". (In
case of an area it is the line which is the decisive "indivisible".)

The line IK above is the ruler. The above plane moves downwards. Then the
two hatched rectangles signify "all the lines", the rectangle KM in the case of the
"straight traverse" and the rectangle KO in case of the "oblique traverse". And these
two collections "all the lines" were considered by Cavalieri to be equal (Fig. 4.1):

$$\mathcal{O}_{KM}(l)_{\text{straight traverse}} = \mathcal{O}_{KO}(l)_{\text{oblique traverse}}$$

Of course, this does not mean that the *areas* of KM and KO are equal. The
reason is that the "traverse" of the "ruler" differs in both cases: in case of KM it is
"straight", but in case of KO it is "oblique". It is eminently important to *compare*
both "traverses" with each other, that is to say, their *ratio*.

Nowadays we describe this ratio with the help of the sine of the angle of
inclination of the rectangle KO. Cavalieri does not do this.

It is crucial that Cavalieri does *not* say, the collection "all the lines" makes up the
area. This would be nonsense. (We will *prove* this below!) Instead, Cavalieri takes
the ratio of two of those collections:

$$\mathcal{O}_{KM}(l) : \mathcal{O}_{KO}(l) \qquad \text{or more general} \qquad \mathcal{O}_{F_1}(l) : \mathcal{O}_{F_2}(l) \,.$$

He compares *only this ratio* to the ratio of the considered areas and that only in
case both areas belong to the same plane. Consequently, the case of the "oblique
traverse" is excluded. Then he has:

$$\mathcal{O}_{F_1}(l) : \mathcal{O}_{F_2}(l) \ = \ F_1 : F_2 \,.$$

It follows that the ratio of the collections "all the lines" of the two surfaces F_1 and
F_2 is the same as the ratio of their areas.

Cavalieri proves this last equality in all detail, i.e. according to Euclid's standards.

We will not go into these details here and just accept the result of Kirsti Andersen's research: by this means Cavalieri succeeded, with all mathematical rigour, in calculating some intricately formed areas. (We quietly pass over the fact that he had to make up his mind anew when faced with differently shaped areas. Leibniz enabled us to deal with this much better.)

Our only aim was to present the principle invented by Cavalieri. The picture shows the uninteresting case where the areas of two rectangles are to be determined. In this case we obviously do not need a new method, for we know: this area is length times width. Cavalieri's method is only of interest if more complicated areas are to be found.

Evangelista Torricelli

Another contemporary of Galilei and Cavalieri is Evangelista Torricelli (1608–47). At first, Torricelli refused to accept Cavalieri's method; but later on he was thrilled by it and thus subscribed to it.

However, Torricelli misunderstood Cavalieri's method. For, contrary to him, he asserted that Cavalieri's strange collections "all the lines" were thought to be *identical to* the area. In other words, Torricelli used just that equation which Cavalieri painstakingly shunned, namely

$$F = \mathcal{O}_F(l) \qquad \text{(banned equation!)}$$

And as Torricelli intensely promulgated this "adopted" method as Cavalieri's—be it out of ignorance or on purpose— he brought Cavalieri into discredit.

Why Are "All the Lines" Not the Area?

That the last equation is nonsense (and, therefore, was rightfully avoided by Cavalieri) can easily be proved.

We take a rectangle $ABCD$ with its diagonal and conclude:

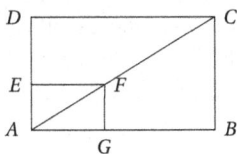

1. Each line EF corresponds to a line FG.
2. Each line EF and each corresponding line FG have the ratio EF : FG, i.e. the ratio AB : BC.
3. But *if* we have $\mathcal{O}_F(l) = F$, *it follows* that the areas of the two large triangles ADC and ABC must have the ratio AB : BC.
4. But they are *obviously* equal and thus we reach a contradiction!

As the first two statements are established facts, the error must arise in the third step. Consequently, the equation $\mathcal{O}_F(l) = F$ must be wrong.

Torricelli's method is of no use! But the reason is not his handling of "indivisibles". (For Cavalieri relied on indivisibles and got a *working* method.) Instead, Torricelli made the *wrong* usage of indivisibles. Wrong means: he identified "all indivisibles" with the "area". Torricelli simply ignored the fact that the continuum does not consist of indivisibles!

Torricelli's mistake disappears, if one does not compose the area from lines but from (very small) pieces of an area instead. The piece $EFGG'F'E'$ consists of two parts with equal areas, divided by FF'. (It is the difference of the larger right angle $AG'F'E'$ and the smaller one $AGFE$, both halved by a diagonal.) Now, if the sides of these parts, EE' and GG', are infinitely small, i.e. if they are taken as indivisibles, then the equality of the two large triangles does follow!

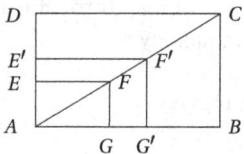

By the way: Torricelli knew about this problem!

Newton's Method of Fluxions

It would be completely unjustified to give an overview of Leibniz' calculus and not to say a word about Isaac Newton (1643–1727), especially as Newton invented his method about ten years prior to Leibniz.

However, Newton's formulation of his method was far less clear than Leibniz'. Besides, it is much more specific than the calculus. Therefore, we will only deal very briefly with Newton's method.

Newton's Method

Newton, too, worked in his manuscripts with "indivisibles". Usually he called them "infinitely small lines" and used them to get new results, as a method of *invention*.

But in case of *proving* his results in his publications he carefully tried to avoid these notions.

An Example

Since 1981, *all* the working papers of Newton have been published, including those which he did not wish to be published. Therefore, everybody has the chance today to witness how he was working: *The Mathematical Papers of Isaac Newton,* volumes 1 to 8.

In the following I present an example from his originally unpublished papers that illustrates how Newton truly worked. To make things easier, I *simplify* Newton's example; but his method is preserved. Newton starts with an equation like

$$x^2 - ax + a^2 = 0.$$

Then he continues: let x be a "fluent" quantity with velocity m. During the infinitely small interval of time o, x will become $x + mo$. (Because length is velocity times duration.) In the equation, $x + mo$ may be substituted in place of x:

$$(x + mo)^2 \quad - a(x + mo) + a^2 =$$
$$x^2 + 2 \cdot x \cdot mo + (mo)^2 - ax - a \cdot mo + a^2 = 0.$$

When the terms of the first equation are erased, we get:

$$2 \cdot x \cdot mo + (mo)^2 - a \cdot mo = 0.$$

Newton divides all by o and gets

$$2 \cdot x \cdot m + m^2 o - a \cdot m = 0.$$

And then he writes boldly:

Since o is supposed to be infinitely small, terms which have it as a factor will be equivalent to nothing in respect to the others. I, therefore, cast them out and there remains

$$2 \cdot x \cdot m - a \cdot m = 0 \quad \text{or more simply:} \quad 2x - a = 0.$$

Every physicist knows: Newton's result is correct. But in regard to mathematical or philosophical standards, Newton's method leaves much to be desired. *At first* he divides by the quantity o—and, therefore, implicitly assumes it to be $\neq 0$— because a division by 0 is forbidden, and *thereafter* he acts as if $o = 0$!

According to mathematics or logic, such reasoning cannot be justified; it is acceptable only by its success.

Nearly one hundred years were to pass, till mathematics succeeded in rendering this crooked reasoning logically sound. Today it is known that Leibniz faced the same problem—and that he solved it unobjectionably with the help of an intricate idea (pp. 32f).

Fluxions and Fluents

Newton calls his variable quantities "fluxions" and "fluents". The *velocity* of the variable x he describes as "fluxion" and sometimes he wrote this fluxion as "$\dot{x}$".

Generally speaking, Newton's *first* problem is to determine the velocity of a known variable quantity. Newton's *second* problem is the reversal of this: to determine the distance covered when the velocity of the moving object is known.

It is easy to express this in Leibniz' own language:

1. If y is given, determine $\frac{dy}{dt}$.
2. If $\frac{dy}{dt}$ is given, determine y.

This is due to the fact that Leibniz *writes down all* the involved quantities. (Just as Descartes then demanded!) Clearly, "velocity" comes along with "time", and, therefore, Leibniz explicitly wrote it down as t. But Newton did not!

Newton's formulae concerning motion do not show time as a variable.

At best, time is *hidden deep* in the notation "$\dot{x}$". Each physicist today can immediately translate this in Leibniz' language:

$$\dot{x} = \frac{dx}{dt}.$$

However, Newton has no knowledge of this language and instead writes "m" (as we have just seen) or "$\dot{x}$".

This reveals the *second* shortcoming of Newton's method: *it is conceptually too restricted.*

Newton's method *demands* that every variable depends on time. But not each problem in the world is of this kind. Sometimes, a variable does not depend on time *alone* but in addition to some other quantity: temperature, pressure, height, etc.

It happens that *one* variable depends on *two* quantities. For Leibniz there is no difficulty: he simply writes *all* the variables of the problem down (as was demanded by Descartes in his particular problems).

Without this approach, Newton gets into severe difficulties: in the case of two variables he has to trace both of them back to time, then to solve the problem and finally, he has to eliminate time again.

This is possible. But it is not simple. It is of little surprise then that British mathematics fell behind in the further development of calculus: Leibniz' notations were more general and thus more applicable.

Literature

Andersen, K. (1985). Cavalieri's method of indivisibles. *Archive for History of Exact Sciences, 31*, 291–367.

Becker, O. (1954). *Grundlagen der Mathematik in geschichtlicher Entwicklung.* Freiburg, München: Karl Alber, [2]1964.

Cavalieri, B. (1647). *Exercitationes geometricae sex.* Bologna: Iacobi Montji. https://www.e-rara. ch/zuz/content/titleinfo/14611669. http://dx.doi.org/10.3931/e-rara-53675

de Gandt, F. (1995). *Force and geometry in Newton's Principia.* Princeton, NJ: Princeton University Press.

Flasch, K. (2001). *Nicolaus Cusanus.* München: C. H. Beck.

Gerhardt, C. I. (Ed.) (1849–1863). *Leibnizens mathematischen Schriften,* 7 vols, Berlin: Weidmannsche Buchhandlung. http://www.archive.org

Leibniz, G. W. (2011). *Die mathematischen Zeitschriftenartikel.* Ed.: Heinz-Jürgen Heß & Malte-Ludolff Babin. Hildesheim, Zürich, New York: Georg Olms Verlag.

von Kues, N. (2002). *Philosophisch-theologische Werke.* Darmstadt: Wissenschaftliche Buchgesellschaft.

Wallner, C. R. (1903). Die Wandlungen des Indivisibilienbegriffs von Cavalieri bis Wallis. *Bibliotheca Mathematica. Zeitschrift für Geschichte der mathematischen Wissenschaften.* third series, 4: pp. 28–47.

Whiteside, D. T. (Ed.) (1967–1981). *The mathematical papers of Isaac Newton* (vols 1–8). Cambridge: Cambridge University Press.

Chapter 5
Do Infinite Numbers Exist?—An Unresolved Dispute Between Leibniz and Johann Bernoulli

A Correspondence

Leibniz and the brothers Jacob and Johann Bernoulli were a dream team. On friendly terms and encouraging each other, they further developed the system first invented by Leibniz and turned it into a productive tool for science and technology. In time, the considerably younger of the brothers, Johann, became Leibniz' closest mathematical friend.

Normally, Johann Bernoulli and Leibniz complemented each other. If one of them made a mistake and the other became aware of it, it went without saying that it was discussed and corrected.

However, there was one big mathematical issue upon which they were unable to come to an agreement. The hotly disputed point was: do infinite numbers exist?

This dispute continued for nine months, from June 1698 until February 1699, alongside various other topics in their correspondence. Nevertheless, these two celebrated mathematical minds never reached unanimity regarding this issue.

Isn't this astounding? Two very well harmonizing mathematicians are unable to settle a mathematical controversy? It is! And, therefore, it is worth taking a closer look at the nature of their dispute.

The Subject of the Controversy

The subject of this friendly controversy was the endless halving of one:

$$\frac{1}{2}, \quad \frac{1}{4}, \quad \frac{1}{8}, \quad \frac{1}{16}, \quad \frac{1}{32}, \quad \cdots \qquad \text{(L)}$$

D. D. Spalt, *A Brief History of Analysis*,
https://doi.org/10.1007/978-3-031-00650-0_5

Today such an infinite object is called a "sequence", in those times it was called a "series". But names do not matter and, therefore, we shall choose the modern term "sequence".

Harmony

It is clear and undisputed between both opponents that:

> This sequence has infinitely many terms.

For, if there were only finitely many terms, there would be a last one. Yet, according to our rule this term is to be halved—and obviously this would be *possible*.

A more sophisticated argument is a *proof by contradiction:* suppose, we only have a finite number of terms. These could be counted, let us say we count to m. Consequently, the last term would be $\frac{1}{2^m}$:

$$\frac{1}{2}, \quad \frac{1}{4}, \quad \frac{1}{8}, \quad \frac{1}{16}, \quad \frac{1}{32}, \quad \cdots \quad \frac{1}{2^m}.$$

According to the rule there would exist a next term:

$$\frac{1}{2}, \quad \frac{1}{4}, \quad \frac{1}{8}, \quad \frac{1}{16}, \quad \frac{1}{32}, \quad \cdots \quad \frac{1}{2^m}, \quad \frac{1}{2^{m+1}},$$

and, therefore, contrary to our assumption, $\frac{1}{2^m}$ would *not* be the last term! That is why our assumption *must* be false.

Johann Bernoulli's Exciting Position

In his letter of **26 August 1698** Johann Bernoulli sets forth the following exciting chain of reasoning:

1. This sequence L *has* infinitely many terms. Therefore, there *exist* infinitely many terms.
2. But if there exist *infinitely many* terms, "infinite" is a number—and consequently there *exists* a *most infinite* term.
3. But if there exists *one* most infinite term then there exist others too.

That is why the sequence of halves is to be written *properly* containing an infinitely large number i:

$$\frac{1}{2}, \quad \frac{1}{4}, \quad \frac{1}{8}, \quad \frac{1}{16}, \quad \frac{1}{32}, \quad \cdots \quad \frac{1}{i}, \quad \frac{1}{2i}, \quad \frac{1}{4i}, \quad \cdots \qquad \text{(B)}$$

A completely new idea! As far as we know, nobody ever thought in this way before!

Johann Bernoulli did not express his idea in this formula, but we have enough evidence to assume that this is what he *thought*. (This can be seen in the way his most prominent pupil, Leonhard Euler, wrote these ideas down much later.—Euler's treatment of the matter will be the subject on p. 77.)

Johann Bernoulli's Prudence

Johann Bernoulli had a good reason *not* to write formula B down, for he knew his friend Leibniz to be a philosophical thinker and, as such convinced that *infinite* "wholes" do not exist. Therefore, he had to be careful in talking to Leibniz about "infinitely large" numbers. ("Whole" means not only an aggregate but also a true object that embodies its parts *in a specific way*.)

Another Shared (Mathematical) Point of View ...

To both it was crystal-clear: for pure mathematical reasons a "largest number" *cannot* exist.

For *if* we could call it i and add 1—whoops, we have a larger number! The *assumption* "there exists a largest number" has led to a contradiction—consequently, the assumption *must* be false: *a* largest number *does not exist*.

We record:

> There exists no largest number.

... with Different, Even Contrary Consequences

From this well established and proven mathematical theorem Leibniz and Johann Bernoulli draw conclusions which are not only *different* but even *completely contrary:*

> Leibniz says: Therefore, infinite numbers *do not* exist *at all.*
> Johann Bernoulli says: infinite numbers *do exist*—however, *not a largest one* but infinitely many, always more.

That means: Leibniz thinks the sequence of all halvings to be (only) L, while for Johann Bernoulli it is *more, namely* B.

Johann Bernoulli's Position in Dispute

Johann Bernoulli was convinced that he could prove his position mathematically. On **8 November** he wrote to Leibniz: "If the infinitely small terms of the sequence do not exist, then not all terms exist." That is to say: for Johann Bernoulli *only* sequence B is right but not sequence L, because L does *not* contain *all* terms.

On **18 November** Leibniz countered: "These infinitely small terms are impossible."

However, that is strong stuff! Leibniz not only claimed to be right (i.e. the sequence under consideration *has* shape L, not B), but moreover that he was able to *prove* that he is right.

Unfortunately, Leibniz did *not disclose* his proof. That's really a pity! For no such proof is known to date. (And I venture: it is impossible.)

Johann Bernoulli Argues

On **6 December** Johann Bernoulli argued for his mathematical viewpoint as follows: "If the terms of the sequence are *numerically* infinite, then necessarily a most infinite term must exist, and it must be infinitely many times smaller than a finite term."

Apart from the final sentence it was the same argument he had already advanced on 26 August.

Leibniz Holds Against

Leibniz did not agree. On **17 December** he called the conclusiveness of Johann Bernoulli's argument into question: "Moreover, I could not say what hinders us to think of a sequence in which each term is finite in size, but which has an infinite number of terms."

Johann Bernoulli Provides the Evidence for His View

On **7 January 1699** Johann Bernoulli formulated his response:

> I can prove this theorem easily in the following way: if we have *ten* terms, there exists necessarily the *tenth*, if *a hundred,* then the *hundredth,* if *a thousand,* then the *thousandth,* and consequently, if we have numerically *infinitely many* terms, then there exists the *most infinite* (infinitesimal) term.

We see: thus Johann Bernoulli *supports* his *pure assertion* by *an argument.* But is his reasoning conclusive?

Leibniz Is Doubtful

Leibniz was not convinced and replies on **23 January 1699:**

> Yet against this, one is allowed to object that in this case the inference from the finite to the infinite is not conclusive, and if one says there exist infinitely many terms that does not mean a certain *number,* instead it is only stated that there exist more than can be expressed by any limited number.

That is to say, Leibniz was convinced: not every multitude is a number and that "infinitely (large)" is no number.

Johann Bernoulli must have been totally flabbergasted: Leibniz frequently reasoned by relying on just this principle, which this time he does not want to accept. We remember the *Law of Continuity* (p. 34), which for Leibniz is of great value and power. Why should this principle not work *in this case?*

Leibniz, possibly anticipating this, confirmed his doubt by the following additional statement:

> In just the same way I could take it upon myself to argue: among ten numbers one is the last, and it is also the greatest of them; consequently, in the totality of quantities there is a last one, and it is the greatest of all, too. Nevertheless, it is my opinion that such a number comprises a contradiction.

That is clever because it obscures the view. Johann Bernoulli's argument was: If there exist n terms, then there exists the nth term. Leibniz plainly *side-steps* this reasoning by saying: the "greatest" term of an infinite sequence is a contradiction. Well, this just avoids Johann Bernoulli's argument, as he speaks about the nth term, not about the "greatest". Thus, Leibniz' reasoning is *correct* but does not apply to Johann Bernoulli's reasoning.

Summary: on 7 January 1699 Johann Bernoulli proved his view, though *not conclusively.* On 23 January 1699 Leibniz argued against Johann Bernoulli, though *not conclusively.*

The End of This Debate: The Disagreement Continues to Exist

Johann Bernoulli's proof is not conclusive—nor is Leibniz' reply. As a consequence, Johann Bernoulli has no objective reason for a reply and does not respond. (Another reason might have been his respect for the older man.)

But Leibniz interpreted the silence in his favour and believed he could confirm his victory. On **21 February 1699** Leibniz wrote:

> You do not answer in response to my argument that the logical inference "if there exist infinitely many terms, consequently, also a most infinite term" is not conclusive.

Did Leibniz really give a reason that "the inference from the finite to the infinite is not conclusive in this case"?—No, he didn't; he only obscured the argument.

It seems as if Leibniz was somewhat aware of the weakness of his position, for he continued:

> I do accept an *infinite multitude*, but this multitude does not constitute a *number* nor a uniform *whole*.
>
>   This means nothing else than the existence of more terms than can be signified by any number, just as there exists the multitude—or the notion—of all numbers; but this multitude itself is neither a number nor a uniform whole.

Leibniz gave *two* arguments: (1) Infinity is not a *number;* (2) and no *whole*.

The *philosophical argument* ("no uniform whole") we will not discuss here.

The *mathematical argument* is this: Leibniz thinks a "number" to be an object which could be given a certain ratio to "one": $\frac{5}{1}$, or $\frac{d}{s}$, where d is the diagonal and $s = 1$ the side of the square. (Or: $\frac{1-\frac{1}{3}+\frac{1}{5}-\frac{1}{7}+-\cdots}{1}$ as on p. 27.) Obviously, infinity *does not fit* this concept of number.

That is the mathematical reason why Leibniz does not accept "infinity" to be a number, and this is plainly correct—according to his way of thinking. And he kept the upper hand in regard to Johann Bernoulli, for his friend did not present another concept of "number" which includes infinity as a number.

We can conclude:

1. The existence or non-existence of "infinite" numbers cannot simply be deduced from the *notion* of "infinity".
2. Leibniz' concept of number does not allow for infinite numbers; they do *not* exist for him.

So, anybody who wants to stand up to Leibniz *has* to offer a new and different notion of "number".

Looking Ahead

As we are yet to see, the concept of number which we use *today* was invented only in 1872, about 200 years after this dispute between Leibniz and Johann Bernoulli (first attempts were made already in 1838 and 1849 and remained largely unnoticed: see below, pp. 201f). Its basic precondition, however, was the *acceptance* of the mathematical "infinite" in some way, at least as a concept of *number*; i.e. that it was mathematically legitimate to accept "infinite" aggregates in mathematics as *well-defined* objects.

This modern state of mathematics, which was reached, say, in 1872, is neither the position of Leibniz nor that of Johann Bernoulli. This new view demands *more* than Leibniz was willing to accept, *viz* to handle "infinite" mathematically legitimized objects. But this new view also demands *less* than Johann Bernoulli presupposed for his mathematics, for it does not create "infinite" natural numbers.

Considering the Real Significance of This Problem: An Inconsistency in the Actual Mathematical Thinking

Decimal Numbers Today: Like Johann Bernoulli Then

This debate is not horribly theoretical; at least not for mathematicians. What is the number $\pi = 3.14159\,26535\,8979\ldots$ about?

We do have a completely precise *definition* of this number: it is the ratio of the semi-circle to the radius of the circle. But do we have it also as a decimal *number?*

This decimal number has infinitely many digits, but they are not determined by a general rule—although they are well-defined, each of them.

Are we consequently and *according to its decimal representation* allowed to accept the number π as a well-defined, *definite* mathematical object?

Leibniz must deny this, while Johann Bernoulli agrees. Leibniz had to accept that each digit of this decimal number *can* be determined—but this *does not* determine the whole number, the aggregate of its digits. Johann Bernoulli thought this to be too sophisticated; in his view (see above) the whole number *is* determined. (We realize: the notion "determined" is not determined clearly and distinctly.)

Nowadays we are used to accepting Johann Bernoulli's view. (Some minorities do not.)

Natural Numbers Today: Like Leibniz Then

However, if we change from the decimal representation of the number π to the totality $\{1, 2, 3, \ldots\}$ of the natural numbers, Johann Bernoulli can find there "infinitely large" numbers too: $\{1, 2, 3, \ldots, i, i + 1, i + 2, \ldots \}$. But Leibniz denies the existence of those passionately.

Yet today, in the case of natural numbers, we stick to Leibniz' view.

Upshot: Anything Goes in Today's Mathematics!

Leibniz and Johann Bernoulli clearly had opposite views concerning the "infinite" numbers: Leibniz denied their existence and even believed them to be "impossible"—whereas Johann Bernoulli took them as self-evident and *easy* to manage.

However, today's mainstream of mathematics casually *embraces* both views:

- The *existence of decimal numbers* and their unrestricted usage is general accepted—which is the mathematical view of Johann Bernoulli. (Smaller circles, e.g. the constructivists, do not agree.)
- Concerning the *natural numbers* matters are opposite: today's mainstream refuses to take "infinite" natural numbers into consideration—just as Leibniz did. (Smaller circles choose the opposite position, last but not least the proponents of nonstandard-analysis.)

The mainstream is the historically grown view. We realize: taken in total, the mathematical thinking of today is not consistent. Not even mathematics! (Of course, the expression "in total" is taken here not in the sense of Leibniz' "whole".)

Literature

Buchenau, A. (1904–06). *Gottfried Wilhelm Leibniz – Hauptschriften zur Philosophie,* 2 vols, ed.: Ernst Cassirer. Vols 107, 108 of the series *Philosophische Bibliothek*, Hamburg: Felix Meiner, ³1966.

Gerhardt, C. I. (Ed.) (1849–1863). *Leibnizens mathematische Schriften.* (7 vols). Berlin: Weidmannsche Buchhandlung. http://www.archive.org

Chapter 6
Johann Bernoulli's Rules for Differentials—What Does "Equal" Mean?

Johann Bernoulli's Rules for Differentials—Part 1: Preparation

Review of Leibniz' Ideas

Earlier we have seen how Leibniz developed the geometrical notion "differential" (pp. 32f). The Leibnizian differential dy is a line segment, where its length is determined as the ratio to a given (finite) line segment with the help of a limiting process (in the figure on p. 33).

This calculation with geometrical objects relies on Leibniz' Law of Continuity. The decisive aspect is to apply this law to *ratios* (or *fractions*) (see calculation on p. 35).

Johann Bernoulli Generalizes

Johann Bernoulli threw off the shackles of Geometry. Descartes was discharged, or more exactly, he was declared outdated. The Leibnizian differential calculus was about *geometrical* variables x, y, dx, dy etc. Johann Bernoulli, in dispensing with Geometry, deals *more generally* with "variable" quantities.

(We have seen that Leibniz had already used geometrical quantities as *variables*: p. 36. With respect to this, Johann Bernoulli offers nothing new.)

From Leibniz' Law of Continuity to Johann Bernoulli's First Postulate

As Johann Bernoulli dispenses with geometrical concepts, he no longer has Leibniz' Law of Continuity (see p. 35) at his disposal. Clearly, he needs a replacement.

And indeed, he has an alternative. He sets down three postulates. The first goes as follows:

> *1. Postulate.*
> A quantity which is increased or decreased by an infinitely smaller quantity is neither increased nor decreased.

A surprise will remain, even if we read very carefully: a quantity which is changed, should *not* be changed? Is there a greater contradiction? Mathematics must be free from contradictions!—What is Johann Bernoulli thinking about?

This is a very hard nut to crack. We need an intermediate stage.

The two other postulates concern geometry, and so we will ignore them.

What Does "Equal" Mean?

The Evident Facts

In mathematics "equal" does not mean "identical" or "the same". Of course not, for otherwise we could not get off the ground: identities (like $4 = 4$ or $a = a$) are of no benefit to mathematics.

> Mathematics exists because of *equations* which are NOT *identities*.
> The mathematical "equal" means, more exactly, "equal in a well-defined respect".

When Descartes writes "$x = 3$" that means: "The root of the equation is 3." That is: the "root of this equation" and the number "3" do relate to each other.— Or possibly: "This line segment has length 3." That is: the "length of this line segment" and the number "3" do relate (in these circumstances).—It depends on the concrete circumstances: are equations to be solved? Are geometrical figures to be constructed? Often the problem is to assign a number to specific geometrical objects (e.g. a line segment) or in algebra (e.g. an unknown of an equation). Or it is about different signifiers for the same thing, which are set "equal", e.g. $(a + b)^2$ and y.

Descartes does not write equations like "$3 = 3$" or "$x = x$", for they are of no use for his mathematics. (In logic, things are different; but logic is not under discussion here.)

What Johann Bernoulli's First Postulate Is All About

Now back to our puzzle: Johann Bernoulli's First Postulate. We may possibly guess what he *means*. It should be read like this:

> If the quantity x is increased or decreased by the quantity dx (i.e. $x \pm dx$) than *this* may be treated—sometimes—like x.

Or more shortly:

> $x \pm dx$ is *in some respect* equal to x.

Remarkable and important is the following: Johann Bernoulli *does* NOT *write this as an equality!* And that is sensible, because such an equation would read:

$$x \pm dx = x. \qquad \text{(forbidden equation!)}$$

Why should this equation be *forbidden?*—Simply, because it implies *conclusively* and according to the general laws of calculation:

$$\pm dx = 0. \qquad \text{(false equation!)}$$

But this *must absolutely be avoided!* For, if $dx = 0$ the entire calculus was to collapse: calculations with added (or subtracted) noughts are fruitless, just like the use of *identities*. The first boils down to the second: if $dx = 0$ it follows $x \pm dx = x \pm 0 = x$, and the First Postulate simply comes down to $x = x$. Nothing of any use.

How This Could Be Written

So far we have understood Johann Bernoulli's First Postulate to be an equation *in a certain sense* ("equal in some respect") and at the same time NOT to be *an equation in the usual sense*.

What are we to do about this?

Since Descartes, mathematics has needed three more centuries for the idea to arise that it is possible to use *several* DIFFERENT *equalities at the same time besides each other* (preliminary Weierstraß: p. 177; then Cantor: p. 195; the exception proves the rule: p. 120).

These equations must of course be distinguished. Especially, it is not permissible to designate the *new, second equality* introduced in Johann Bernoulli's First Postulate by the old "=". This we just examined.

> *Consequently,* we choose *another* equality sign.

Which one? In principle the shape of the sign does not matter. We could write "=*", or "=$_2$", or "≈", or anything else.—Let's agree on "≈". Consequently, from now on we shall not write "$x \pm dx = x$" (for this is *wrong!*) but

$$x \pm dx \approx x$$

in Johann Bernoulli's First Postulate. Then we will get no problems. (We may read this as: "is approximately equal".)

> Johann Bernoulli's *First Postulate* is, in modern notation:
> $$x \pm dx \approx x.$$

For us this is not a huge step, but it was within mathematics of the time and, indeed, well beyond: it took about three centuries for this success. (Naturally, there were hidden forerunners: see p. 120.)

What Is This Huge Step About?

Descartes tought us the *pure formal equation*. Mathematics of the twentieth century (more precisely: structural mathematics—the Bourbaki group and its followers) brought us the idea of using *several distinct* signs for equality besides each other. Consequently we have been able to write really intricate equations like

$$(x + dx)^2 - x^2 = x^2 + 2x\,dx + dx^2 - x^2$$
$$= 2x\,dx + dx^2 = (2x + dx)\,dx$$
$$= (x + [x + dx])\,dx$$
$$\approx (x + x)\,dx$$
$$= 2x\,dx\,.$$

For the transition from line three to four we used Johann Bernoulli's First Postulate—visible through the *other* equality sign; and only there.

We record:

> Mathematics developed in such a way in the twentieth century that *more* (i.e. different) equality relations could be used *together.*

Those *other* "equalities" (besides =) are called "equivalences"; instead of talking about "other equations" one speaks, more pompously, about "equivalence relations". A foreign term does not make things better or easier. At face value, an equivalence relation is *another* equality, another relation of the form "equal in respect to ...".

The Equalities Must Be Consistent

If one is to calculate with both equalities together, as we just did, those must be consistent. (We should have checked this already!) Consistency in this case means: one of the equalities *must* imply the other. And only one! This is met here:

> If $a = b$ then also $a \approx b$. (Of course, the opposite is not valid!)

In this case, the second of these two "equals" is called "finer" than the other: $\approx$ is "finer" than $=$.—And naturally, $=$ is called "coarser" than $\approx$.

Johann Bernoulli's Rules for Differentials—Part 2: Execution

The power of the "differentials" is: their calculation is subject to *laws*. Oddly enough, these laws are called "rules".

Leibniz had already laid down these rules and proved them. Of course, he argued in geometric terms, and he did this meticulously. We passed over this procedure.

Instead, we look at Johann Bernoulli: which were his Rules for Differentials? And how did he prove them?

Rules 1 and 2: Addition and Subtraction

Johann Bernoulli writes:

Rule 1
The differential of a sum is the sum of the differentials of each single summand.

His proof goes as follows:

The differential of the quantity $a + x$ is dx, because *a denotes a "constant"*, as is assumed now and in the following.

To wit, if $a + 0$ and $x + e$ are added, the sum is $a + x + e$; if the subtrahend $a + x$ is deducted, the remainder becomes $e = dx$. Q. E. D.

(in both meanings of the phrase: which *was to have been* proved and that a statement which was to be proved, finally *had been* proved).

There is not much to say about this proof. Because a is assumed to be a "constant" it does not change, and consequently $da = 0$. On the contrary x changes: $dx \neq 0$ (for otherwise all would be nothing: $d(a + x) = da + dx = 0 + 0 = 0$).

Regarding our detour to the twentieth century we notice one aspect: Johann Bernoulli does not write *any* equality sign, except at the end, where it is completely unnecessary.

Today we are used to writing this proof as a sequence of equalities, and we only need the common equal sign (we do *not need* the *First Postulate:*)

$$d(a + x) = [(a + da) + (x + dx)] - [a + x] = da + dx = 0 + dx = dx .$$

The same idea applies if *both* summands are "variables": $d(x + y) = dx + dy$. This does not need to be elaborated here.

It is the same with subtraction: $d(x - y) = dx - dy$.

Rules 3 and 4: Multiplication and Division

We have already treated the special case $d(x^2) = (x + dx)^2 - x^2$ above, where we used *two different* equality signs! Let us simply add: Johann Bernoulli presents yet again a proof *without writing one single formal equation*, except, as before, at the end, where it is completely superfluous.

Thereafter Johann Bernoulli shows: the differential of x^3 is equal to $3 \cdot x^2 \cdot dx$; again without writing any equation.

Johann Bernoulli proves the general case next, where he, for the first time, does a sloppy job, because he actually writes the result as an equation:

$$d(x^p) = px^{p-1} dx .$$ (false equation!)

Correct would be:

$$d(x^p) \approx px^{p-1} dx ,$$

but this was not part of mathematics during the eighteenth (or even during the nineteenth) century.

When dealing with the rule for division, interestingly Johann Bernoulli writes the decisive step as follows:

$$\frac{ey - fx}{yy + fy} = \text{(following Postulate 1)} \; \frac{ey - fx}{yy} = \frac{y\,dx - x\,dy}{yy}$$

(that is to say: besides $e = dx$ he has $f = dy$). Today we would write this:

$$\frac{dx \cdot y - dy \cdot x}{y \cdot y + dy \cdot y} \approx \frac{dx \cdot y - dy \cdot x}{y \cdot y} = \frac{y \cdot dx - x \cdot dy}{y \cdot y}.$$

For our "$\approx$", Johann Bernoulli wrote "= (following Postulate 1)" and thus he actually made use of *another sign for equality!*

Rule 5: Roots

Finally, Johann Bernoulli deduces his Rule 5 for roots. Written correctly, it is:

$$d \left(\sqrt[p]{x} \right) \approx \frac{dx}{p \sqrt[p]{x^{p-1}}} \, .$$

Needless to say: Johann Bernoulli writes again the incorrect sign "=".

The First Book Containing the Rules for Differentials Stems from de l'Hospital

Johann Bernoulli did not publish the "Rules for Differentials" himself but taught the French nobleman Marquis de de l'Hospital. (Of course, he was paid for his teaching.) As a consequence of this, this nobleman, in 1696, printed the first textbook on the differential calculus: *Analyse des infiniment petits pour l'intelligence des lignes courbes*; under his own name, because he was the author. However, at least in the preface he named Johann Bernoulli for having been his teacher. *Naturally*, de l'Hospital was only interested in geometry (curves) and not in algebra—the latter being Johann Bernoulli's invention.

(Actually, in the early 1920s a copy of a manuscript was found in the library of Basel University which most certainly stems from Johann Bernoulli's lecture notes for de l'Hospital.)

The Marquis de de l'Hospital was an eminently astute scholar. In his book the Rules for Differentials are scrupulously presented and proved—and that without the use of equality signs! Even when his teacher Johann Bernoulli was careless in his manuscript (as it was not written for being printed!) de l'Hospital argues correctly. This does him credit.

A Precursor of de l'Hospital's Book!

The story above is not complete(ly correct). One year earlier than de l'Hospital's book, in the year 1695, a book by Bernard Nieuwentijt had appeared in Amsterdam: *Analysis infinitorum,* i.e. "Analysis of the infinite". In its last chapter (starting on p. 278) this author presents Leibniz' Rules for Differentials, too. He had come across them in the *Acta eruditorum*, the first article in which Leibniz was to present his invention in 1684 (p. 32).

Especially interesting is the following: Nieuwentijt does not only state these rules but also justifies them, whereas Leibniz in his article had given *no* justifications of his rules! Nieuwentijt was thus forced to find such justifications by himself. The problem: Nieuwentijt's way of thinking was fundamentally different from that of Leibniz.

An Unsuitable Justification of the Rules for Differentials

How does Nieuwentijt prove the product rule then? At first, Nieuwentijt proceeds in the usual manner: like Johann Bernoulli (both followed the English Isaac Barrow who invented this idea), he calls the increase of x shortly e and that of y shortly f and states:

$$(x + e)(y + f) - xy = xf + ye + ef .$$

In the last expression Nieuwentijt simply omitted ef—and arrived at the product rule:

$$d(xy) = x \, dy + y \, dx .$$

What was his explanation for simply ignoring ef? He argued completely differently from Leibniz!

In the very first definition of his book, Nieuwentijt states the principle of a strong empiricist philosophy: he *solely* took such quantities as given which do not have a magnitude that exceeds human imagination. Consequently, for Nieuwentijt those quantities which are too small for human imagination are just nought.

In Leibniz, Nieuwentijt reads that the differentials are not nought. But if the differentials are diminishing *further*—as it is the case when they are multiplied: $ef = dx \cdot dy$—Nieuwentijt decided this product to be *necessarily* zero. All this follows Nieuwentijt's philosophy and proofs given at the start of his book; he simply refers to that.—End of Nieuwentijt's proof of the product rule.

Nieuwentijt endeavoured to create a kind of mathematics which included only such quantities which do not exceed *human imagination*. But a thinking like this cannot cope with differentials.

Actually, single differentials are already "infinitely small" quantities. For Leibniz this meant: they get *smaller than any given quantity*. That is why it is arguable whether or not differentials fit to "the limits of human intelligence" at all.

If not, they have to be taken to be nought—and then there is no differential calculus. But in case they (just about) fit within the "limits of human imagination", undeniably their products (and other, even smaller quantities) are nought—and again a differential calculus like that of Leibniz becomes impossible.

History is, as often, full of surprises and contradictions: the first book of Leibniz' Rules for Differentials, together with their justifications, was presented within a system of notions that renders it absurd. Luckily, de l'Hospital's book appeared the following year and offered a reasonable and accurate presentation.

Literature

Leibniz, G. W. (1684). Nova methodus pro maximis et minimis, itemque tangentibus, quae nec fractas, nec irrationales quantitates moratur, & singulare pro illis calculi genus. *Acta eruditorum*, pp. 467–473. cited after: Leibniz 2011.

Leibniz, G. W. (2011). *Die mathematischen Zeitschriftenartikel.* Ed.: Heinz-Jürgen Heß & Malte-Ludolff Babin. Hildesheim, Zürich, New York: Georg Olms Verlag.

Marquis de l'Hospital 1696. *Analyse des infiniment petits pour l'intelligence des lignes courbes.* Paris: François Montalant, [2]1716. http://gallica.bnf.fr/ark:/12148/bpt6k205444w, http://www.archive.org/details/infinimentpetits1716lhos00uoft

Nieuwentijt, B. (1695). *Analysis infinitorum, seu curvilineorum proprietates ex polygonorum natura deductae.* Amsterdam: Joannem Wolters. http://dx.doi.org/10.3931/e-rara-10853

Schafheitlin, P. (1924). *Die Differentialrechnung von Johann Bernoulli aus dem Jahre 1691/92, Vorlesungen über Differentialrechnung.* Leipzig: Akademische Verlagsgesellschaft.

Chapter 7
Euler and the Absolute Reign of Formal Calculation

The Absolute Monarch of Eighteenth Century Mathematics

Leonhard Euler (1707–83) dominated mathematics during the eighteenth century. The royal homes of this enduring absolute sovereign of mathematics were the scientific Academy in St. Petersburg (the new capital of the newly restructured Russian state)—from 1731 till 1741 and again from 1766—and in between, the Berlin Academy. (It had been Leibniz' suggestion to cover Europe with a network of scientific academies.) Euler never gave a lecture at university.

Leonhard Euler left behind the most comprehensive collection of mathematical writings known today. They are documented in more than a hundred large sized, bulky volumes. This is the more astounding as he went completely blind in 1766. Subsequently he just dictated his treatises. This shows an exceptional mathematician endowed with an exceptional memory.

From the age of 13, Euler studied mathematics for five years at the university of Basel under the tutelage of Johann Bernoulli—and this besides his main subject of theology. The very busy Johann Bernoulli recommended certain works to his student Euler, but found no time to lecture him privately. However, on Saturdays Euler was allowed to put questions to Johann Bernoulli. Being very respectful, Euler always tried to ask as little as possible in order not to bother Johann Bernoulli unnecessarily.

The Invention of the Principal Notion of Analysis: "Function"

> Euler made *function* the principal notion of analysis.

Leibniz had already used the word "function", but not in a specific, well-defined way (see p. 40). Only Johann Bernoulli provided a precise meaning for this word.

D. D. Spalt, *A Brief History of Analysis*,
https://doi.org/10.1007/978-3-031-00650-0_7

In a treatise, printed in 1718 (i.e. two years before Euler started being supervised by him) he stated:

> One calls *function* of a variable quantity any quantity which is arbitrarily composed from variable quantities and constants.

Johann Bernoulli had detached the foundational concepts of Leibniz' theory from geometry. Instead, he dealt more generally with "variable quantities" like "abscissas", "ordinates" etc. (see p. 59). Consequently, he built from these "quantities" a new concept: the "function". The "quantities" are to *compose* a "function".

His disciple Leonhard Euler accepted this idea. In the year 1748 Euler's *Introductio in analysin infinitorum* (*Introduction to the analysis of the infinite*) appeared, the first textbook on analysis. For quite a few decades it set the standards for the entire field of analysis.

The Components of Functions: Quantities

What Is a Quantity?

Johann Bernoulli did not disclose what a non-geometrical "quantity" ought to be. Euler is hardly more informative. Casually, Euler once came up with the statement that "each quantity can be increased or decreased endlessly", thereby expressing his conviction that this is due to the *nature* of a quantity. (By the way, this definition is not contained in his textbook on analysis but in that of the differential calculus, which appeared in 1755.)

This example shows: Euler was not a gifted philosopher. The citation names *properties* of a "quantity" but does not state, *what* a quantity ought to be exactly.

The First Kinds of Quantities. Euler's Characterization of Quantities Is Insufficient

Consequently, we do not know *what* Euler took "quantities" to be. What we do know is: Euler conceived a "quantity" to be something which can be "increased or decreased endlessly".

All the more surprised we then read the very first definition of his textbook of analysis:

> A CONSTANT *quantity is a definite quantity which always keeps the same value.*

The concepts "quantity" and "constant quantity" are *definitions* of Euler. Logically, a "constant quantity" is a "quantity". But Euler, contrarily, demands that a "constant quantity" should lack exactly those properties which he named as a *char-*

acterization of a "quantity"—*viz* that it can be "increased or diminished endlessly". Euler is caught in a logical error: there should be no "constant quantities".

After all, Euler set these concepts himself. Therefore we are left with the somewhat unfriendly conclusion:

> With the help of his notation "constant quantity" Euler circumnavigates his failure to define a general concept of quantity.

However, our criticism is of no help for our understanding of Euler. Euler just ignored this fundamental flaw of his concepts and argued as if it did not exist. We are committed to check whether this defect of Euler's concepts has any (mathematical) consequences. (It does not seem so.)

A "constant quantity" always retains "the same value". We would rather not ask Euler what a "value" ought to be. Instead, we will confine ourselves to the observation of what Euler used as values. First of all he takes the "numbers." But there is more to come—we should take care!

The Second Kind of Quantities

Besides the "constant quantities" (which are those that retain "always the same value") there exists a second kind of quantities. This is stated by Euler in the second definition of his textbook of analysis:

> A VARIABLE QUANTITY *is an indefinite or general quantity which can take any definite value whatsoever.*

Consequently, "variable" means for Euler: *indefinite* or *general*. Euler takes the "variable" quantity as the *general* quantity. Consequently the "constant" (or: "definite") quantity is for him a *special* quantity. To sum up: a "variable" quantity takes *all* the "constant" (or: "definite") quantities.

Following Descartes, Euler denotes the variable quantities with the last letters of the alphabet, sometimes written small: z, y, x etc., sometimes not: Z, Y, X, etc. (see p. 75).

Euler's Algebraic Concept of Function

Now we turn to Euler's definition of that notion of analysis, which has been *central* ever since:

> A *function* of a variable quantity is a variable quantity which is described (represented) by an *analytical expression* which is in some way or other composed from the variable quantity and numbers or from constant quantities.

To be meticulous: the above is not the *exact* formulation of Euler, but if one strives for philosophical precision it amounts to that.

Instead of "analytical expression", following a proposal from Rüdiger Thiele, one can more precisely say: "calculatory expression" ("Rechenausdruck"). Its meaning is what was invented by Descartes: a formula composed from *letters* (in twofold signification: definite or indefinite quantities), *numbers* and *signs of calculation.* Examples for such expressions are:

$$a + 3z; \quad az - 4zz; \quad az + b\sqrt{aa - 4zz}; \quad c^z \quad \text{etc.}$$

These calculatory expressions are *not* yet the "functions" but only *denote, represent* them. The "calculatory expression" is Euler's specification of what his teacher Johann Bernoulli called "composed from".

Simple but Important Consequences from Euler's Notion of Function

Let us notice three simple but important consequences to be drawn from Euler's notion of function.

1. *Whether* a calculatory expression *really* denotes a "function" has to be checked in each particular case. For example: if we let $c = 1$ in the calculatory expression c^z we get 1^z. But 1^z is always 1 whatever the value of the variable quantity z will be. *Consequently,* the calculatory expression 1^z is not a "variable" quantity but a "definite" one, and therefore *no* function! The calculatory expression 1^z does *not* designate a "function" (but a "definite" quantity). Similarly for the calculatory expression z^0: whichever value we assign to z, we always get $z^0 = 1$. (Except for the case $z = 0$ for we have $0^0 = 0$—but even *an exception from the rule does not matter.*) That is why Euler does not take 1^z and z^0 as functions but as constant quantities—We conclude:

   > Following Euler, *not every* calculatory expression denotes a function!

2. Euler *explicitly demands* that a "variable" quantity takes "all definite values whatsoever". To illustrate this, he explains the calculatory expression in detail

$$\sqrt{9 - zz}.$$

As under the square root sign we have something not greater than 9, this calculatory expression seems to denote numbers not greater than 3. But this consideration is wrong! The "variable" quantity is to take "all definite values whatsoever"—and for Euler this means: *all imaginable* numbers; and therefore also, e.g. the (not "actual" but) at least *imaginable* number $4\sqrt{-1}$! Yet, if you take $z = 4\sqrt{-1}$, you get

$$\sqrt{9 - zz} = \sqrt{9 - \left[4\sqrt{-1}\right]^2} = \sqrt{9 - 4^2 \cdot \sqrt{-1}^2} = \sqrt{9 - (-16)} = \sqrt{25} = 5 > 3.$$

$\sqrt{9 - z^2} = u$ is equivalent to $9 = u^2 + z^2$, which is an expression where z and u have the same status. As z shall denote *all* "definite values"—and takes even all "imaginable" numbers—, u shall do so, too. The function which is denoted as $u = \sqrt{9 - z^2}$ thus takes *all* numbers, which are "imaginable" according to Euler.

Today, numbers with the component $\sqrt{-1}$ are called "complex" numbers. As Euler *expressly* demands to include those numbers in the domain for his functions, we can state *in modern terminology:*

> In principle, Euler admits complex functions.

3. Finally, a very important and for us surprising view:

> Euler does not distinguish between *finite* and *infinite* calculatory expressions.

Surely a surprise—but if inspected more closely, not unfounded. No doubt, it would be possible to consider the number calculated by Leibniz (p. 23):

$$1 - \tfrac{1}{3} + \tfrac{1}{5} - \tfrac{1}{7} + \tfrac{1}{9} - + \ldots = \tfrac{\pi}{4}$$

to be something special, but this peculiarity is *not solely* founded on the infinite character of the calculatory expression. This is shown by the calculatory expression on the left which is also infinite

$$1 + \tfrac{1}{2} + \tfrac{1}{4} + \tfrac{1}{8} + \ldots = 2,$$

for it equals a completely common and in no sense peculiar number.

The equation

$$1 + x + x^2 + x^3 + x^4 + \ldots = \frac{1}{1 - x}, \qquad (*)$$

shows even more clearly that a distinction between *finite* and *infinite* calculatory expressions is meaningless.

This infinite series emerges from the usual procedure of division if applied to the *calculatory expression* instead of two *numbers,* i.e. akin to *long algebraic division:*

$$
\begin{array}{r}
1 \\
\hline
1 - x \,)\, 1 \\
1 - x \\
\hline
x
\end{array}
$$

In detail: if we divide 1 by the difference $1 - x$ we at first obtain 1 with the rest x (because from 1 there is the product $1 \cdot (1 - x)$ to be subtracted: $1 - (1 - x) = x$):

$$1 : (1 - x) = 1 + \frac{x}{1 - x} .$$

Check:

$$1 + \frac{x}{1 - x} = \frac{1 - x}{1 - x} + \frac{x}{1 - x} = \frac{1}{1 - x}$$

and consequently the equation before is correct. Continuing this process we get:

$$\frac{1}{1 - x} = 1 + \frac{x}{1 - x}$$

$$\frac{1}{1 - x} = 1 + x + \frac{x^2}{1 - x}$$

$$\frac{1}{1 - x} = 1 + x + x^2 + \frac{x^3}{1 - x}$$

$$\frac{1}{1 - x} = 1 + x + x^2 + x^3 + \frac{x^4}{1 - x}$$

etc.

How Did Euler Denote Functions?

At the beginning but usually also later, Euler denotes a general "function" of a "variable" quantity x (or y or z) by the respective capital letter: X (or Y or Z). For example, he writes "$Z = \sqrt{9 - zz}$".

Sometimes he uses the letter "f" and appends the variable quantity: for Euler, "fx" indicates a function of the variable quantity x. In these cases Euler does not write brackets, as we do today—we write "$f(x)$". Such brackets are used by Euler only if the respective variable quantity is *composed*, e.g. if it is a sum of two quantities: "$f(x + y)$".

> For Euler "x" denotes a variable quantity.

A Standard Form for Functions

Euler proves in his differential calculus that *each* function of a variable quantity can be presented according to the general scheme:

$$b + X(x - a) \; - \frac{1}{2}P(x - a)^2 + \frac{1}{2 \cdot 3}Q(x - a)^3 \; - \frac{1}{2 \cdot 3 \cdot 4}R(x - a)^4$$

$$+ \frac{1}{2 \cdot 3 \cdot 4 \cdot 5}S(x - a)^5 \; - + \text{etc.}$$

where a is any definite (finite) value; and also b, X, P, Q, R, S etc. are definite values which have to be calculated by means of differential calculus from the given function.

(We realize that Euler is not consistent in his notations, for *strictly speaking* the capital letters X, P, Q, R, S should designate *functions,* whereas in this case they designate *values.*)

For example b is the value of the given function for $x = a$ (as all other summands are then equal to zero). X is the value of what we call today the "first derivative" of the given function, again at the value $x = a$. (If you know differential calculus, you see this at once.) Today, this is called the "Taylor series" of the given function.

In particular cases this sum can have a finite number of terms.

Our Problems with This Theorem of Euler

From today's point of view we have some objections to the validity of this theorem. One can say: in some cases (i.e. for some values a) the equation is false!

But such a judgement does not hold for Euler. For Euler it is clear (as in the previous case, where he does not count z^0 to be a "function"): *the exception proves the rule!* Euler's theorems are valid *in general,* for his "variable" quantities are "indefinite" or "general" quantities. *In particular cases* there might be an exception.

> Euler's analysis focuses on "variables" (not on "values"), and, consequently, so do his theorems.

A Daring Calculation with Infinite Numbers

Euler, in his textbook on analysis from 1748, offers a completely surprising kind of calculation. Its *result* is still undisputed to this day. However, the manner in which Euler reaches his result is judged questionable, problematic, or even wrong in today's mathematics. Let's take a look!

From the Powers of Ten to the Exponential Quantity

The way we usually write numbers is founded on the powers of 10; the decimal system. It is the reason why we can write every number in *standard index form*. The *value* of a digit is determined by its place, its distance from and location to the decimal point.

$$3; \quad 30; \quad 300; \quad \ldots; \quad \quad 0.3; \quad 0.03; \quad \ldots$$

This is how the exponential quantity

$$10^z .$$

works. The larger the value of the exponent z the larger is the value of the exponential quantity.

Needless to say, Euler took the most general perspective and represented the number 10 by some "constant" quantity a:

$$a^z .$$

We limit ourselves to the analogy of a to 10 and take only $a > 1$.

The Exponential Function

Now we turn to Euler's startling way of dealing with the exponential function. He says:

1. We have $a^0 = 1$.
2. If the exponent increases, so does the value of the number.
3. If the exponent only increases *very little*—more precisely: only "infinitely little"—then the value of the number also increases only *very little,* more precisely: only "infinitely little".
   Euler writes this in a very clever way:

$$a^\omega = 1 + \psi .$$

ω as well as ψ denote two infinitely small numbers. Next, Euler connects them:
4. $\psi = k \cdot \omega$.
   Consequently, he gets

$$a^\omega = 1 + k \cdot \omega .$$

The arbitrary exponent (z) of a^z is chosen to be an infinitely small value ($z = \omega$); and a fundamental insight into the nature of the exponential function is expressed by the *form* of its value, i.e. by $1 + k \cdot \omega$.

Finally, we need one more idea and the rest will simply be mathematical routine.

5. We have to return from the infinitely small exponent ω to the finite exponent z. This we achieve by multiplying ω by an *infinitely large* number i: $z = \omega \cdot i$. Because of $(a^\omega)^i = a^{\omega \cdot i}$, we get

$$a^z = a^{\omega \cdot i} = (1 + k \cdot \omega)^i .$$

The rest is mathematical routine. On the right we have a formula which we can develop according to the Binomial Theorem as an infinite series:

$$a^{\omega i} = 1 + \frac{i}{1}k\omega + \frac{i(i-1)}{1 \cdot 2}k^2\omega^2 + \frac{i(i-1)(i-2)}{1 \cdot 2 \cdot 3}k^3\omega^3 + \cdots$$

Through rearranging $z = \omega \cdot i$, we obtain $\omega = \frac{z}{i}$ and through cancelling in each fraction the i (e.g. in the second summand: $\frac{i}{1}k \cdot \frac{z}{i} = \frac{1}{1}kz$ etc.), we have

$$a^z = 1 + \frac{1}{1}kz + \frac{1(i-1)}{1 \cdot 2i}k^2z^2 + \frac{1(i-1)(i-2)}{1 \cdot 2i \cdot 3i}k^3z^3$$
$$+ \frac{1(i-1)(i-2)(i-3)}{1 \cdot 2i \cdot 3i \cdot 4i}k^4z^4 + \cdots$$

And then comes the final clue, Euler's *trick of cancellation:* each fraction has the infinitely large number i as often in the numerator as in its denominator. As $i - 1$ is nearly equal to i (for i is infinitely large) and the same with $i - 2$ etc., Euler keeps cancelling i against $i - 1$, i against $i - 2$ and so forth, and only the simplified expression

$$a^z = 1 + \frac{kz}{1} + \frac{k^2z^2}{1 \cdot 2} + \frac{k^3z^3}{1 \cdot 2 \cdot 3} + \frac{k^4z^4}{1 \cdot 2 \cdot 3 \cdot 4} + \cdots$$

remains. If we chose a special value for k, we fix the value of a. The a which results from $k = 1$ we often call "Euler's number" today and denote it by "e". So we get

$$e^z = 1 + \frac{z}{1} + \frac{z^2}{1 \cdot 2} + \frac{z^3}{1 \cdot 2 \cdot 3} + \frac{z^4}{1 \cdot 2 \cdot 3 \cdot 4} + \cdots$$

Even the mathematical layman might find this comprehensible: this infinite series does *not* cause any problem. More precisely, it has for any value of z (clearly, we only speak about *finite* values!) a definite value which can, at least in principle, be calculated. For, the denominators of the fractions on the right increase very quickly and will surpass any value of z *whatsoever.* Once the denominators "overtake" the numerators, the fractions will decrease *rapidly* and will become arbitrarily small. Since Leibniz we know: quantities with "arbitrarily small" changes are alright, because we may calculate them to any degree of accuracy.

Let me add a technical detail. Unfortunately, the "rapidly" above is important! If the summands decrease too slowly—take e.g. $1+\frac{1}{2}+\frac{1}{3}+\frac{1}{4}+\frac{1}{5}+\ldots$—then this will *not* be enough to provide the series with a finite value. The considerations above are *not sufficiently* precise: it matters how rapidly the remaining terms diminish.

(Surely, the sum of the n terms $\frac{1}{n+1}+\frac{1}{n+2}+\ldots+\frac{1}{2n}$ will be greater as if we only take the smallest, i.e. the last one, n times—and consequently greater than $n\cdot\frac{1}{2n}$; on p. 86 we calculate this in more detail.) However, the latter is $\frac{1}{2}$! In other words: if we proceed in the harmonic series $1+\frac{1}{2}+\frac{1}{3}+\frac{1}{4}+\frac{1}{5}+\ldots$ far enough, the terms *again and again* add up to more than $\frac{1}{2}$; therefore the sum of this series is *greater than* $1+\frac{1}{2}+\frac{1}{2}+\ldots$ and, consequently, not finite.)

The Ingenious Trick—Or: Euler's Cheat

However, in Euler's calculation there is a bold step! Euler cancels i against $i-1$; than this *and i* against $i-2$; than all this *and i* against $i-3$, and so forth.

But this is *false!* For, it equates (a) $i-1$ and i, which means: $-1=0$; (b) and *in addition* $i-2$ and i, which means: $-2=0$; (c) and *just as much* $-3=0$ and so forth! These equations are not only false but they also contradict each other!

Well, you may say: that is true—yet these errors are *infinitely small!* Today, we may write this in the following way (as i is infinitely large):

$$i-1\approx i, \quad i-2\approx i, \quad i-3\approx i \quad \text{etc.}$$

and all is well.

By no means! For Euler has to concede: his trick of reduction produces an error in *every* fraction—although an infinitely small one. Nevertheless, there are infinitely many fractions and, consequently, *infinitely many such errors!* And the integral calculus teaches how to add infinitely many infinitely small amounts to a *finite* value (we followed Leibniz' calculation on pp. 29f).

Outcome: Euler's calculation is not false. (Here we must trust the mathematical specialists: if one assigns any value to x, the values on *both* sides of Euler's equation completely coincide.) However, the equation *is not conclusively proved.* An argument is missing that *in this case* infinitely many infinitely small errors do *not* add up to a finite amount.

Today's readers may take comfort from the following. Let $k=1$. The general term $\alpha_n(z)=\frac{1(i-1)(i-2)\cdots(i-(n-1))}{1\cdot2i\cdots ni}z^n$ is smaller than the general term $\beta_n(z)=\frac{z^n}{n!}$ of the e^z-series, taken absolute values. Though, the last series is known to converge—take the ratio test. Consequently, also the series of $\alpha_n(z)$ converges for finite z, let's say to $A(z)$. As both series are converging we get for finite $\varepsilon > 0$ a finite m_0 such that if $m > m_0$ we have: $\left| \sum_0^m \alpha_n(z) - A(z) \right| < \varepsilon$ as well as $\left| \sum_0^m \beta_n(z) - e^z \right| < \varepsilon$.

However, both partial sums $\sum_0^m$ having only finitely many summands are (like their summands) infinitely close and, consequently, their difference is certainly $< \varepsilon$. Hence $|A(z) - e^z| < 3\varepsilon$ and therefore it follows that Euler's error is indeed infinitely small: $A(z) \approx e^z$. Nevertheless, remember: *Euler* does not have ANY concept of "convergence" in his textbook on analysis!

Euler's Concepts of Numbers

Leonhard Euler performed an inconceivable amount of calculations (by hand!), surely more than any other mathematician has ever done. He calculated with numbers as well as with formulas. One of his calculations with formulas we have just seen. We shall not present any of his calculations with numbers, for this is not part of any consideration of the foundational *concepts* of analysis, which are at stake here.

It has already been stated that Euler was not a gifted philosopher. Therefore, we should not be astonished that Euler keeps silent in regard of the essence of number and its concept. This topic he handles pragmatically. He is not afraid to *found* mathematical laws by recourse to everyday life. (Philosophically this is unacceptable.) For example, he justifies the multiplication rules for negative numbers (among them: "minus times plus is minus") by interpreting negative numbers as *debt*. The most difficult of these rules ("minus times minus is plus") he just *decrees:* "I say, the opposite must be the answer."—as minus times plus is already minus.

The "irrational numbers" Euler presents plainly under the aspect of required calculations, as *roots:* $\sqrt{2}$, $\sqrt{3}$, $\sqrt{5}$ (irregularities of this kind do not bother him at all) etc. Besides, there exist of course $\sqrt[3]{2}$, $\sqrt[3]{3}$ etc., clearly also $\sqrt[4]{2}$ ... and so on and so forth. A huge zoo, without a general view or theory. For the sake of calculating that will do.

Similarly, $\sqrt{-a}$ may be subjected to the usual rules of calculation without further ado. The arithmetic genius and "calculator", Euler, is amply satisfied with this.

Euler's treatment of the exponential function has shown that he also uses *infinite* (natural) numbers. Euler's teacher Johann Bernoulli was reluctant, at least in his exchange with Leibniz, to denote such numbers *in formulas* and confined himself to mention them *only verbally* (pp. 53f). Euler has no such qualms—and was also able to *show* which fruitful mathematical results could be gained with the help of formulas.

Infinitely large natural numbers are plainly *objects* of mathematical thinking which cannot easily be *characterized* clearly and distinctly. This we have seen in the dispute between Leibniz and Johann Bernoulli (pp. 52f).

Euler's often happy-go-lucky, really careless, presentation of "infinite" numbers is therefore completely disastrous and plainly awful. So he actually writes:

> Consequently, one can justly say that 1 divided by 0 indicates an infinitely large number or ∞.

And, shortly after, he goes even further and writes:

> Because $\frac{1}{0}$ indicates an infinitely great number and, undoubtedly, $\frac{2}{0}$ is two times as great, it is obvious that even an infinitely great number may be enlarged 2 times.

His *intention* is obvious: $2i > i$ ("*i*" for Latin "infinitus"). However, not even Euler should get away with such an inconsistent way of thinking like $\frac{1}{0} < \frac{2}{0}$, because when taking reciprocals we inevitably get nonsense: $\frac{0}{1} > \frac{0}{2}$ or $0 > 0$, which is absolutely forbidden in mathematics.

Even the greatest geniuses have their weaknesses! Thus we are left with the following judgement:

> *Euler did not deliver any contribution to the clarification of the notion of number.*

Anyway, one positive outcome we get from this completely crazy passage in Euler: obviously, besides the "numbers" there exists something else for him that stands in as a "value"—and that is ∞. Thus, let us finish this consideration with the optimistic result:

> Euler counts, besides the "numbers", also ∞ as a "value".

Analysis Without Continuity and Convergence

Hopefully, today's mathematicians are wondering about the fact that Euler's analysis gets along without those two concepts which are of the greatest importance in modern analysis: *continuity* and *convergence*. Although nobody will have read *every* textbook of analysis in use, we can be fairly sure: each of them deals with these two notions. Yet, Euler's textbook does not.

This is one of the rather rare cases when the question "why?" can be answered easily and clearly. The question reads: why does Euler get along in his analysis without the notions of *continuity* and *convergence?* Answer: because both concepts, *convergence* and *continuity,* deal with mathematical facts which are about *values*— nonetheless, singular *values* are not the subjects of Euler's analysis; they are considered *only in their entirety.*

We remember: Euler demands from a "variable" quantity to take "all definite values without exception" (p. 71). His analysis deals at most with a "constant" quantity (and, of course, with "numbers") as "values;" but not as today, with values as *constituent* and *foundational* notions.

> Euler's analysis ignores the notion of "value."

Euler delivers no arguments for that.

With this, his analysis is *fundamentally* simpler than today's analysis. It lacks these two notions, "continuity" and "convergence"—and they are both, each in itself, somewhat intricate. "Continuity" and "convergence" are not concepts suitable for small talk. "Continuity" and "convergence" are really difficult technical terms. These logically somewhat tricky notions were only theorized by the mathematicians of the nineteenth century. We will trace this story later (pp. 103f).

To sum up Euler's analysis: its *only* central concept is "function". It is based on numbers and the concept of "quantity". The last exists in twofold manner: as "constant" and as "variable".

In the nineteenth century, the concepts "continuity" and "convergence" were introduced and became more and more fruitful. That is why we finally ask what Euler thought about these concepts. For yes, he himself actually *used* them!

Continuity According to Euler

The "calculatory expression" is the central notion in Euler's analysis. Therefore, it makes sense to transfer this concept to other mathematical fields, e.g. to geometry. Euler did this in the following way:

The principal object of geometry in the eighteenth century is no longer—as once for Euclid—the straight line, but, far more generally, the "curved line." Descartes invented the pure formula and showed how to denote certain straight lines ("abscissas" and "ordinates") by "x" and "y" and from these, together with the signs denoting arithmetic operations, he built a "formula". Leibniz made these "x" and "y" fluid. He overthrew Descartes' decree to understand them as definite "numbers" and, instead, took them as "variable quantities".

Subsequently Euler fused both, Descartes' formula and the then principal notion of geometry, the "curved lines". He decided: the *proper* curved lines are those which can be described by a "formula." As the formula, more precisely, the "calculatory expression", is Euler's favourite concept, he introduced the definition:

A "curved line" is called *"continuous"* if it can be described by a calculatory expression.

An example is: $(x + 2)^2 + 3$.

In Euler's conceptual world this means: a "continuous" curved line is the *geometrical image* of a "function", for a "calculatory expression" describes in Euler's eyes a "function" which is therefore its *algebraic form.*

Without being explicitly stated, Euler's emphasis on this definition of "continuity" is that it must be *one single* calculatory expression. As we already stressed, Euler demands a "variable" (that is the "x" in the formula) to take "all definite

Fig. 7.1 Example 1

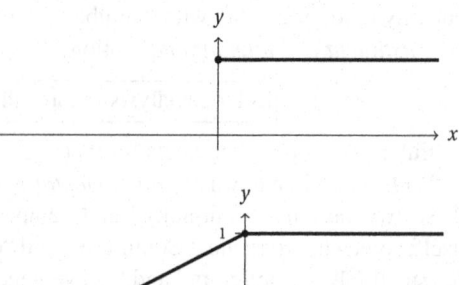

Fig. 7.2 Example 2

values without exception". Consequently, it was for Euler *unthinkable* to supply a *"function"* with a RESTRICTED domain of values and to say, e.g.: "take the function $y = 2x + 1$ for all values of x, starting from $x = 3$." or even: "take $y = 1$ for all $x \geq 0$—by which we nowadays describe the half-line" (Fig. 7.1).

Similarly, we describe the following broken straight line (Fig. 7.2)

by

$$y = \begin{cases} \frac{1}{2}(x + 2) & \text{if } -2 \leq x < 0 \\ 1 & \text{if } \quad 0 \leq x, \end{cases}$$

and we are using *two* "calculatory expressions":

$$\frac{1}{2}(x + 2) \text{ and } 1.$$

For us the picture shows the geometric image of a "function"—and more: the image of a *(in today's understanding)* "continuous" function!—Euler would not have understood this. For him, *not one* of the two examples can be described by "one calculatory expression" and therefore they are not "continuous" *curved lines* for him.

And Euler is right! None of these lines can be described by *one single* calculatory expression; additional *particular distinctions* are needed, e.g. (in)equalities or a *principal change of the notion of function* which renders it possible to implement the method of "Fourier analysis". However, according to the notions of his own analysis, Euler is absolutely right: *according to his own concept of function* the lines given in these examples are not "continuous".

Euler's *Second* Notion of Function

As only some curved lines can be described by a single calculatory expression, it is impossible for Euler to deal with all curved lines in this way. Of course, he was aware of this.

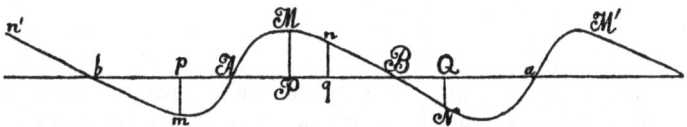

Fig. 7.3 Euler's second notion of function (1748/1759)

Nevertheless, Euler is the last to feel restricted because of mathematical concepts. On the contrary, whenever he is faced with the problem of describing *any* "curved lines" mathematically—e.g. those of Fig. 7.3—, he changes his view. In these cases Euler does not fall back on his teacher Johann Bernoulli's notion of function as a calculatory expression. Instead, he describes the "curved line" in just the same way as Leibniz had already done: he expresses the *dependency* of the "ordinate" on the "abscissa". This way, Euler specified a *second* concept of function and even created a special notion for it:

> As it is generally possible to represent the ordinate of some curved line by $f : z$, where z is the abscissa, we take $\mathcal{A} \, \mathcal{M} \, \mathcal{B}$ to be the curved line, where the ordinates $\mathcal{P} \, \mathcal{M}$ are the *functions*[1] of the abscissas $\mathcal{A} \, \mathcal{P}$ and denoted with help of the sign $f : z$, the result of which is $\mathcal{P} \mathcal{M} = f : (t \sqrt{b})$.

More clearly: the respective "abscissa" z, e.g. $\mathcal{A} \, \mathcal{P}$, refers to the "ordinate" $\mathcal{P} \mathcal{M}$. These "ordinates", altogether, constitute the "geometrical" function; given as formula: $\mathcal{P} \, \mathcal{M} = f : (z)$.

Of course, this way Euler is out of the woods. For he is *now* able to deal with *each* "curved line" and make it a "function" *in this sense*, denoted by "$f : (\cdot)$". (The "ordinates" are represented through their *end points: n', b, m, $\mathcal{A}$, $\mathcal{M}$, n, $\mathcal{B}$, $\mathcal{N}$, a, $\mathcal{M}'$.)

This approach—to resort to a second notion of function depending on the problem—is unusual. Of course, scrupulous readers of Euler became aware of this and set out to study Euler's *second (geometrical)* concept of function, as for instance the Alsatian mathematician Louis Arbogast (1759–1803) in his treatise *Dissertation on the nature of arbitrary functions, which enter into the integrals of partial differential equations* (original in French, in the year 1791). Later this fell into oblivion and today's books on the history of analysis do not mention it. (Only Prof Dr Thomas Sonar in his book with the pompous, but completely misleading title *3000 years of analysis* (in German), hints at it, but unfortunately he forgot to name his source.)

---

[1] Emphasis added.

Outlook

Nowadays the concept of continuity is fundamental to analysis. It is a notion of analysis, not of geometry. Nonetheless, Euler's concept of continuity refers to "curved lines" and is therefore a property of *geometrical* entities. In other words:

> The concept of "continuity" which is common today, *must* differ from that which Euler invented.

Its definition and its development will be shown later (pp. 107f, pp.125f).

Convergence According to Euler

The two notions of "convergence" and "sum" (of a series, more precisely "true sum") are for Euler strongly related. We will explain them separately.

Initially, the notion of "convergence" refers to *infinite* sums, also called "series". We remember one of Leibniz' problems (p. 23):

$$1 - \tfrac{1}{3} + \tfrac{1}{5} - \tfrac{1}{7} + \tfrac{1}{9} - + \ldots$$

In the case of this infinite sum Leibniz had shown by way of a completely accurate proof that it has a finite value. We may say: he proved that it "converges."

Euler came up with a spectacular deduction of the *exponential series*

$$e^z = 1 + \tfrac{z}{1} + \tfrac{z^2}{1 \cdot 2} + \tfrac{z^3}{1 \cdot 2 \cdot 3} + \tfrac{z^4}{1 \cdot 2 \cdot 3 \cdot 4} + \cdots$$

(p. 77). Indeed, this series always gives a finite value, whichever (finite) value z might take. (And we have also seen that an infinite series *not necessarily* has a *finite* sum.)

Nowadays we say that a series is "convergent" only if it has a *finite* sum, otherwise it is not.

Convergence and Divergence

For Euler, things are different! Euler, too, coined a notion of *convergence*—but one which differs from the one we use today. In a treatise titled *On divergent series* Euler presents the following definition:

> We call a series *convergent* if its terms keep getting smaller and finally disappear completely, like with
>
> $$1 + \tfrac{1}{2} + \tfrac{1}{4} + \tfrac{1}{8} + \tfrac{1}{16} + \tfrac{1}{32} + \text{etc.,}$$

whose sum is indubitably $= 2$.

For, the more terms one adds, the more one approaches the 2. If one adds just a hundred terms, the error to the two is already extremely small: it is a fraction with numerator 1 and a denominator with thirty digits.

Unfortunately, this definition is not as clear as it should be. Euler seems to argue that it is important that a series, which "converges" according to his definition must have a finite sum, too. *But this is not the case!*

One has to read his words *very carefully.* If so, one realizes:

1. It consists of *two* sentences.
2. The first sentence begins with an *explanation,* the *definition;* which is
3. immediately followed by an *example,* introduced by the words "like with".
4. The second sentence refers exclusively to the given *example.*

Consequently, the end of the first sentence together with *the second sentence* do not relate to the definition of "convergence" any more. They are a *consequence* of it, or properly speaking: an *example* thereof.

In short, Euler's definition reads:

We call series "convergent" if their terms continually decrease and finally disappear completely.

That this painstaking interpretation leads to the correct result becomes immediately clear when we read Euler's definition of the opposite of "convergence", namely his definition of "divergence":

Series are called *divergent* if their terms do not tend to nothing, but either never decrease below any given limit or increase to infinity. Such are

$$1 + 1 + 1 + 1 + 1 + \text{etc.}$$

as well as

$$1 + 2 + 3 + 4 + 5 + 6 + \text{etc.},$$

wherein many terms added give an always greater sum.

Again, Euler comes up immediately with examples. However, this time the first sentence is unambiguous and clearly separated from the second one.

To memorize the difference between Euler's "convergence" and ours, we state:

In Euler's conceptual world the harmonic series

$$1 + \tfrac{1}{2} + \tfrac{1}{3} + \tfrac{1}{4} + \tfrac{1}{5} + \ldots$$

is "convergent", although it has no finite sum!

In more detail: whatever natural number k one proposes, it will always be possible to designate a number p which gives:

$$1 + \tfrac{1}{2} + \tfrac{1}{3} + \tfrac{1}{4} + \tfrac{1}{5} + \ldots + \tfrac{1}{p-2} + \tfrac{1}{p-1} + \tfrac{1}{p} > k \ .$$

Even if we have $k = 10^{100}$ this will be possible—although this number very likely exceeds the numbers of the atoms in the universe! And the same will be true for $k = 10^{1000}$, a number which is clearly beyond good and evil.

As already explained on p. 78, the decisive point is that you will *always* find another section of the series which adds up to more than $\tfrac{1}{2}$:

$$\tfrac{1}{n+1} + \tfrac{1}{n+2} + \ldots + \tfrac{1}{2n} > n \cdot \tfrac{1}{2n} = \tfrac{1}{2} \ .$$

And $\tfrac{1}{2}$, added often enough (namely $2k$ times), increases to any given natural number k (for: $2k \cdot \tfrac{1}{2} = k$)! For the p above you may choose the number $2^{2(k-1)}$ to be on the safe side. Realization: the number of the terms to be added, increases exponentially, like 2^x. Notwithstanding, one *always* obtains the fraction $\tfrac{1}{2}$—isn't it strange? And moreover, the particular summands decrease below any given limit—*actually* an entirely astonishing wonder of calculation!

The True Sum

Euler is an outright *consequent formalist!* Of course, he knows about the significance of the *sum of the terms of a series,* especially in the case of a "convergent" series. In case of the series $1 + \tfrac{1}{2} + \tfrac{1}{3} + \tfrac{1}{4} + \tfrac{1}{5} + \ldots$ this sum is infinite, for which Euler writes "∞".

But Euler is not satisfied with that! Referring to the example of the series $*$ from p. 73 and the associated calculations, Euler introduces in his *Differential calculus* the notion of *"true sum":*

> We will avoid these difficulties and apparent contradictions completely, if we give the word *sum* another, unconventional meaning. So we will call the expression which is developed to give rise to the infinite series, the *true*[2] *sum of this series.*

That means: the "true sum" of the series $1 + x + x^2 + x^3 + \ldots$ is $\tfrac{1}{1-x}$. *And that is completely independent of the value of x!* Also for the case $x = 1$ (as for Euler holds $\tfrac{1}{0} = \infty$), and even for $x = 2$:

$$1 + 2 + 2^2 + 2^3 + 2^4 + \ldots = 1 + 2 + 4 + 8 + 16 + \ldots = \tfrac{1}{1-2} = \tfrac{1}{-1} = -1 \ .$$

According to Euler, the formula triumphs over common arithmetic! He even accepts the "true sum" of infinitely many numbers, which increase beyond any limit, to be -1. *It seems impossible to show a greater degree of disdain for numbers and common arithmetic!*

---

[2] This word added, for *evidently* Euler forgot it here.

Of course, in being a human *"calculator"*, Euler takes the "common" sum of $1+2+2^2+2^3+\ldots$ to be nothing else than ∞. However, his *theory* called "Algebraic Analysis" is something very different. It does *not* deal with common arithmetic, with the "common" sum—but with the *"true" sum!*

For Euler the *truth* is in the *formula,* in the *calculatory expression*

The cited formulation from the *Differential Calculus* is not a one-off but it is Euler's *true mathematical conviction.* Twenty years later he repeated this view in its entirety in his *Complete Manual to Algebra* and explained it, this time in the German language by referring to the "remainders" $\frac{x^{n+1}}{1-x}$ that originate from the procedure akin to long division (see p. 74):[3]

At the first glance this looks unfitting.
But we should notice that if one wants to stop somewhere in the above series, one needs always to put an additional fraction there as well.
Thus, if we stop for instance at 64, we have to attach to

$$1 + 2 + 4 + 8 + 16 + 32 + 64$$

also this fraction: $\frac{128}{1-2}$ that is $\frac{128}{-1} = -128$, wherefrom $127 - 128$ emerges, which is -1.
If one moves on without end the fraction falls away, but one also never stands still.

To illustrate this further, let's take a second example for a "true sum" in Euler's sense:

$$1 - 2x + 3x^2 - 4x^3 + 5x^4 - + \ldots = \frac{1}{(1+x)^2} \qquad (\dagger)$$

The "true sum" of the series on the left is the calculatory expression on the right.
(Those who are used to some university maths will easily deduce this result with the help of the method of undetermined coefficients—others will readily accept this without proof; maybe they could just check the result?)
If we have the calculatory expression, it is not too difficult to write down the corresponding power series. However, there is no strategy for the reverse process: *to find the "true sum" of a series is always a mathematical problem.*
Euler faces a problem, which nowadays we confront only in case of "converging" series: even if we have proved the convergence—in today's meaning!—of a series, we do not know its value by a long way! For Euler it is just the same: it is not easy to find the "true sum" of a series, i.e. the corresponding calculatory expression. (This is quite independent of the "convergence" of the series, be it in ours or in Euler's meaning.) Today, we are used to calculating the true "value" of a "converging" series, whereas Euler does not offer a theory about such particular entities as "values." He deals with algebra, with the general: with the "formula."
But one thing seems clear: with his concept of "true sum" Euler *for once* prefers a finite calculatory expression to an infinite one. (Of course, it would be

---

[3] as clumsy and old-style as possible.

truly surprising to find a *universal* principle among the rules created by the highly pragmatic Euler.)

To Sum up Euler's Algebraic Analysis

Euler's concept of analysis was called "Algebraic Analysis". This name fits very well. Our summing up of his theory is this:

> The courtly etiquette of the absolutist royal ruler corresponds in mathematics to Euler's calculatory expression, the formula. Both etiquette and formula, outperform common-sensical behaviour and calculations, respectively—be the outcome in the particular case as absurd or irrational as humanly possible. Both are done according to *fixed and rigid rules* and that at any cost, even if *the obvious facts* speak against it.

(Whoever wants to learn the details about the courtly etiquette in absolutist monarchies may consult the authoritative work of Norbert Elias *The Court Society*, written in 1969, or more recently the book *Das Europa der Könige* from Leonhard Horowski.)

Literature

Arbogast, L. F. A. (1791). *Mémoire sur la nature des fonctions arbitraires qui entrent dans les intégrales des équations aux différentielles partielles*. St. Pétersbourg: Académie Impériale des Sciences.

Bernoulli, J. (1991). Op. CIII. „Remarques sur ce qu'on a donné jusqu'ici de solutions des Problêmes sur les isoperimetres, avec une nouvelle methode courte & facile de les resoudre sans calcul, laquelle s'étend aussi à d'autres problêmes qui ont rapport à ceux-là". [Goldstine (1991). *Die Streitschriften von Jacob und Johann Bernoulli*. Ed.: David Speiser. *Die gesammelten Werke der Mathematiker und Physiker der Familie Bernoulli* (pp. 527–568). Basel, Boston, Berlin: Birkhäuser Verlag].

Elias, N. (1969). *Die höfische Gesellschaft*. Vol. 423 of the series *Suhrkamp Taschenbuch Wissenschaft* (Frankfurt am Main: Suhrkamp, 1981).

Euler, L. (1788a). *Leonhard Eulers Einleitung in die Analysis des Unendlichen*. Erstes Buch. (Berlin: Carl Matzdorf). http://books.google.de/books?id=VwE3AAAAMAAJ&pg=PP2&ots= LNs0o_YK04&lr=, German: Johann Andreas Christian Michelsen.

Euler, (1788b). *Leonhard Eulers Einleitung in die Analysis des Unendlichen*. Zweytes Buch. (Berlin: Carl Matzdorf). http://books.google.de/books?hl=de&lr=&id=EwI3AAAAMAAJ& oi=fnd&pg=PA5&ots=zSybL4AxZo&sig=G-Nioin1m_1c5p6-cGI91jpldWs. German: Johann Andreas Christian Michelsen.

Euler, L. (1828, 1829, 1830). *Leonhard Euler's vollständige Anleitung zur Integralrechnung*, 3 vols (Wien: Carl Gerold), German: Joseph Salomon.

Euler, L. (1836). *Leonhard Eulers Einleitung in die Analysis des Unendlichen*. second volume, new unchanged corrected edition. (Berlin: G. Reimer), German: Johann Andreas Christian Michelsen.

Euler, L. (1983). *Einleitung in die Analysis des Unendlichen von Leonhard Euler*. (Berlin: Springer, new edition: Berlin etc: Springer). http://gdz.sub.uni-goettingen.de/no_cache/dms/load/toc/? IDDOC%=264689, German: H. Maser 1885.

Euler, L. (1745/1748). E 102, Introductio in analysin infinitorum. Tomus secundus. In: A. Speiser (Ed.), *Leonhardi Euleri Opera Omnia.* Vol. 9 of series I (Zürich; Leipzig, Berlin: Orell Füsli; B. G. Teubner, 1945). http://dx.doi.org/10.3931/e-rara-8740, Deutsch: [Euler, L. (1836). *Leonhard Eulers Einleitung in die Analysis des Unendlichen.* second volume, new unchanged corrected edition. (Berlin: G. Reimer), German: Johann Andreas Christian Michelsen].

Euler, L. (1746/1760). E 247, De seriebus divergentibus. In: C. Boehm & G. Faber (Ed.), *Leonhardi Euleri Opera Omnia,* vol 14 of series I, pp. 585–617 (Leipzig, Berlin: B. G. Teubner, 1925). http://math.dartmouth.edu/~euler/

Euler, L. (1748/1750). E 140, Sur la vibration les cordes. In: F. Stüssi (Ed.), *Leonhardi Euleri Opera Omnia,* vol 10 of series II, pp. 63–77 (Leipzig, Berlin: B. G. Teubner, 1947). http://math.dartmouth.edu/~euler/

Euler, L. (1763/1768). E 342, Institutiones calculi integralis. Vol. 1. In: F. Engel & L. Schlesinger (Eds.), *Leonhardi Euleri Opera Omnia.* Vol. 11 of series I (Leipzig, Berlin: B. G. Teubner, 1923). http://urn:nbn:de:bvb:12-bsb10053432-6

Euler, L. (1768/1769). E 387, Vollständige Anleitung zur Algebra. In: Heinrich Weber (Ed.), *Leonhardi Euleri Opera Omnia.* Vol. 1 of series I (Leipzig, Berlin: B. G. Teubner, 1911). http://math.dartmouth.edu/~euler/

Euler, L. (1790–1793). *Leonhard Euler's vollständige Anleitung zur Differenzial-Rechung.* 3 vols, from Latin translated into German and annotated. (Berlin, Libau: Lagarde und Friedrich, 1790, 1790, 1793). http://dx.doi.org/10.3931/e-rara-8624, German: Johann Andreas Christian Michelsen

Goldstine (1991). *Die Streitschriften von Jacob und Johann Bernoulli.* Ed.: David Speiser. *Die gesammelten Werke der Mathematiker und Physiker der Familie Bernoulli* (Basel, Boston, Berlin: Birkhäuser).

Horowswki, L. (2017). *Das Europa der Könige.* (Reinbek bei Hamburg: Rowohlt).

Laugwitz, D. (1986). *Zahlen und Kontinuum. Eine Einführung in die Infinitesimalmathematik.* (Mannheim: Bibliographisches Institut).

Sonar, T. (2011). *3000 Jahre Analysis.* Heidelberg, Dordrecht, London, New York: Springer.

Spalt, D. D. (2011). Welche Funktionsbestimmungen gab Leonhard Euler? *Historia Mathematica, 38,* 485–505. https://doi.org/10.1016/j.hm.2011.05.001

Thiele, R. (1982). *Leonhard Euler.* Vol. 56 of the series *Biographien hervorragender Naturwissenschaftler, Techniker und Mediziner.* Leipzig: BSB Teubner Verlagsgesellschaft.

Chapter 8
Emphases in Algebraic Analysis After Euler

d'Alembert: Philosophical Legitimation of Algebraic Analysis as Well as His Critique of Euler's Concept of Function

Jean-Baptiste le Rond d'Alembert (1717–83), another first-rate mathematician, was a slightly younger contemporary of Euler. He was also active as a philosopher and involved in politics. His name, together with that of Denis Diderot (1713–84) personifies the French *Encyclopédie*, published in the years 1751–80, a work which greatly influenced the French Enlightenment. (However, d'Alembert left the scientific editorial staff of the *Encyclopédie* as early as 1758.)

d'Alembert's Reflections on the Notion of Quantity

In volume 7 of the *Encyclopédie*, published in 1757, d'Alembert starts his entry about *Quantity* with the following sentence: "It is one of those words, the entire world believes to have a clear idea of, but which is nevertheless very difficult to define precisely."

d'Alembert's Critique

D'Alembert starts with the very same notion of quantity which we have already found in Euler: quantity is, what "can be increased or decreased without end".

© The Author(s), under exclusive license to Springer Nature Switzerland AG 2022
D. D. Spalt, *A Brief History of Analysis*,
https://doi.org/10.1007/978-3-031-00650-0_8

At first, d'Alembert explains the importance of the "or" in this definition. Would it instead read "and", both zero and infinity would not meet the requirement of the definition: for zero cannot not be decreased and infinity cannot be increased. However, d'Alembert thinks it cannot be denied that both are "quantities".

We realize: d'Alembert's idea of the infinite is not like Euler's or Johann Bernoulli's, but more like Leibniz'—because Euler as well as Johann Bernoulli allows himself to increase the infinite (i): $i + 1$, $i + 2$, ...; in opposition to this, Leibniz and d'Alembert do not.

d'Alembert's Notion of Quantity

D'Alembert switches abruptly to his own understanding of "quantity" and declares: "It appears to me that *quantity* can be defined well as being something which is composed of parts."

This brings d'Alembert back to Euclid, who had declared two thousand years earlier: "A *number* is a multitude composed of units." And: "A number is *a part* of a number, the less of the greater, when it measures the greater."

And *by the way,* the philosopher d'Alembert now explains the foundational legitimacy of this Algebraic Analysis:

> The quantity[1] exists in each finite being and it expresses itself in an indefinite number which can only be known and understood with the help of a comparison and in relation to another homogene *magnitude.*

Descartes' universal unity is dismissed and the Law of Homogeneity restored, albeit with a new meaning.

Assessment: d'Alembert's Philosophical Legitimation of Algebraic Analysis

Unfortunately we cannot know d'Alembert's *intention* behind this formulation. Maybe he did not want to say something different from what umpteen generations of mathematicians have said already before him: only the fixing of a ruler (the "unit") allows the measuring of a quantity. However, closer inspection of the matter (which includes the historical moment of d'Alembert's writing) gives rise to read this sentence as the *philosophical legitimation of Algebraic Analysis:*

> It is not about *definite* values, but about *all possible* values, without any exception; only that counts.

This was Euler's demand, at the beginning of his textbook (p. 79).

---

[1] d'Alembert first uses "quantité" for "quantity" but in the end he uses "grandeur".

d'Alembert's Critique of Euler's Notion of Function

For many years Euler and d'Alembert competed for the best mathematical description of the vibrating string. Unable to agree—each of them thought his own answer to be superior. As so often in everyday life: they talked past each other. Neither of them understood the other. They had different ideas of the notion of a function.

What Euler understood by "function" *in this case,* was shown on p. 83, but what was d'Alembert's idea?

D'Alembert did not have Johann Bernoulli as his teacher (as Euler did) and was therefore further removed from the geometrical foundations of analysis. D'Alembert stuck to the algebraic notion of function, given by Euler: a function is described by a "calculatory expression". Nonetheless, it was clear to d'Alembert that this was not enough to deal with all the "curved lines" encountered in practical problems—but only with the nice ones: those which Euler had called continuous (p. 81).

D'Alembert tried to liberalize this algebraic notion of function in two steps. His principal idea was to replace the *one* calculatory expression by *two* of them.

1. At first d'Alembert allowed a "function" to be determined by an *equation,* that is to say by two calculatory expressions, related by an equal sign.
   Unfortunately this idea is of no great help. If we have an unknown on *both* sides of the equation it is very doubtful whether we will be able to make y the subject of the equation. (Since the nineteenth century it has been known that this can already fail with equations of the fifth degree. During the eighteenth century one was far more optimistic.) If one does not succeed in isolating the unknown? How should the "function" be further inspected, if it is not given explicitly?

2. D'Alembert tried to modify the notion of function after 1761 and from 1780 he became more specific.
   His writing is not terribly clear, but two points can be discerned:

   a. It is clear that he connects two functions and their "equations" with adjoining domains to make up one "continuous" (because of the *two* expressions!) function. That is to say, he changes his original notion of function decisively by *limiting* its *scope.* For Euler this was *unthinkable!*
   b. He carries over the notion of "continuous" from "curved line" to "function". Contrary to modern concepts, d'Alembert calls a *compound* "function" "continuous", if it describes a *single* "curve" and if these two *different* "functions" at their meeting point (i) take the same "value" and (ii) have the same (one-sided) derived function (increase or decrease of the tangent).

Thus, d'Alembert aims to give the notion of function more flexibility *within the conceptual realm of Algebraic Analysis,* because the algebraic notion of function as demanded by Euler is not very useful in practical circumstances.

d'Alembert's Impulse: Condorcet

If one digs even further into the mathematical sources of the late eighteenth century, more precisely into the treatise *On the continuity of arbitrary functions* of Marie Jean Antoine Nicolas de Caritat, Marquis de Condorcet (1743–94) from 1774, one gets some evidence that d'Alembert's last step was inspired by a much younger contemporary with that illustrious name. However, such subtleties exceed the scope of this book.

Lagrange: Making Algebra the Sole Foundation of Analysis

The last first-rate mathematician who completely subscribed to the spirit of Euler's analysis was Joseph Louis Lagrange (1736–1813). His textbook was published in 1797, the "year V" following the calendar of the revolution, and in 1813 in a second edition. A German translation by Johann Philipp Grüson (1768–1857) appeared as soon as 1798 under the title (in German) *Theory of analytic functions, wherein the principles of differential calculus are given, independently from considerations of the infinitely small or vanishing quantities, the limits or fluxions, and grounded in Algebraic Analysis*. Eventually this text led to the establishment of "Algebraic Analysis" as a title for the new theory. It is an apt name.

Lagrange's New Foundation of Analysis: The Base

The lengthy title of the work encapsulates its content and plan. Lagrange aims to present the theory of functions and the differential calculus as simply as possible—and, consequently, *independently of* such complicated notions as "infinitely small" quantities, "limit" or (as with Newton) "fluxion". Analysis as easy as possible, a commendable plan.

Lagrange's idea is very consequential and can be understood in three steps.

First Step. Lagrange starts at the very beginning, with Euler's notion of function. Accordingly to Euler, a "function" is a "variable" quantity, which is described by a calculatory expression. Lagrange radicalizes this and says:

> A *function* is a calculatory expression.

Following this, *any* calculatory expression is a "function" for Lagrange—be it the description of a variable quantity or not.

Here, he differs from Euler. Euler only considers a calculatory expression as a "function" which describe a *changing* quantity. Lagrange no longer cares whether the calculatory expression can take on only one or more values. For him *every* calculatory expression is a "function".

Let's recapitulate Euler's notion of "calculatory expression".

1. The simplest calculatory expressions are the *sums*, finite ones like $2 + 3x - 5x^2$ or infinite ones like $1 - x + x^2 - x^3 + x^4 - + \ldots$

2. However, a simple fraction can also be written as an (infinite) sum (see formula $*$ on p. 73):

$$\frac{1}{1-x} = 1 + x + x^2 + x^3 + \ldots$$

or (see formula † on p. 87)

$$\frac{1}{(1+x)^2} = 1 - 2x + 3x^2 - 4x^3 + 5x^4 - + \ldots$$

3. Therefore Euler is convinced that *every* function can be described as such an (infinite) *sum*. This he noted in § 59 of his textbook on differential calculus:

> Thereupon it should be beyond doubt that every function can be transformed into such an expression that runs towards the infinite
>
> $$Ax^\alpha + Bx^\beta + Cx^\gamma + Dx^\delta + \cdots, \qquad (\ddagger)$$
>
> wherein the exponents $\alpha, \beta, \gamma, \delta \ldots$ stand for any numbers whatsoever.

Euler chooses Greek letters (alpha, beta, gamma, delta) as exponents. They can also be *arbitrary* numbers.

The Idea of Lagrange

To implement his plan, Lagrange—in his *second step*—seizes on a fact which was originally advanced by Johann Bernoulli (in the year 1694) and again, very clearly and repeatedly, by Euler. In his *Integral Calculus*, Euler restated a theorem, which he had already proven in his *Differential Calculus* (and earlier, see p. 75):

Theorem. If y denotes a function of x, which changes to b, if $z = a$, and if we put $\frac{dy}{dz} = P$, $\frac{dP}{dz} = Q$, $\frac{dQ}{dz} = R$, $\frac{dR}{dz} = S$ etc., then we obtain the general expression:

$$y = b + P(z-a) - \tfrac{1}{2}Q(z-a)^2 + \tfrac{1}{6}R(z-a)^3$$
$$- \tfrac{1}{24}S(z-a)^4 + \tfrac{1}{120}T(z-a)^5 - \text{etc.}$$

For a better understanding of the main idea we simplify this calculatory expression. If $(z - a)$ is replaced by x and the differential quotients P, Q, R, etc. are shortened to $1P = p, -\tfrac{1}{2}Q = q, \tfrac{1}{6}R = r, -\tfrac{1}{24}S = s, \ldots$ we get the following condensed expression:

$$f(a + x) = f(a) + p \cdot x + q \cdot x^2 + r \cdot x^3 + s \cdot x^4 + t \cdot x^5 + \text{etc.} \tag{§}$$

Euler had shown that *each* function, each calculatory expression can be represented in this way. The left side of the equation is new, instead of "$f(x)$" we now have "$f(a + x)$".

It is now Lagrange's idea—his *third step*—to prove that the scheme ‡ is nothing other than Eq. §. In other words, Lagrange proves the following theorem:

> Theorem. EACH *function* $f(x)$ *can be written as a calculatory expression of the form §.*

The proof of this theorem is Lagrange's opening to his textbook on analysis.

If Lagrange were really able to prove this theorem, he would stage a coup d'état. Because of Euler's theorem which we have cited above, one *then* was able to deduce that the coefficient p of the second summand in § is the first differential quotient of the function (*viz* $\frac{dy}{dz}$); the coefficient q of the third summand on the right in § is the second differential quotient (in Euler: $\frac{dP}{dz}$) inclusive of the factor $\frac{1}{1 \cdot 2}$ and the sign; the coefficient r of the forth summand on the right in § is the third differential quotient of the function ($\frac{dQ}{dz}$) inclusive of the factor $\frac{1}{1 \cdot 2 \cdot 3}$; etc. In other words: if Lagrange can prove this theorem he would be able to obtain for every function all differential quotients. *In this case, the differential calculus is founded all at once and entirely without the use of the commonly applied notions like "infinitely small" quantities* etc.

Lagrange introduces also a new notation which is still used today. Instead of the "first differential quotient" $\frac{dy}{dz}$ of the function f Lagrange simply writes "f'"; instead of the second differential quotient $\frac{dP}{dz}$ he writes "f''" etc. Until today, f', f'' etc. are called the "first", "second", ... "derivatives" of the function f.

A Contemporary Criticism on Lagrange's Plan

August Leopold Crelle (1780–1855), who does not count as a first-rate mathematician, edited Lagrange's textbook in a new German translation in 1823. Crelle saw fit to launch a fundamental criticism of Lagrange. He wrote:

> In my opinion, the proof that the series expansion of an arbitrary function $f(a + x)$ only contains positive integer powers of the quantity x, is firstly defective or at least very weak and much too complicated to found the principles of a whole science; and, secondly, I think such a proof is completely *superfluous*.

That is strong stuff, and the latter judgement is very noteworthy: why is the proof of this theorem in Crelle's eyes "superfluous"?

Crelle makes it easy for himself and says: later in the book it is shown, how it works—but if you show *how* it works, you need not prove *that it works*.

This is obviously self-deception. Of course, Lagrange does not show in his book how to expand *all* functions in a series: the book is finite (the original has 296 pages), but there are infinitely many functions! And of course, Lagrange (without any doubt being a first-rate mathematician) did not resort to such kind of argument. He was sure to have a watertight, general proof for his theorem.

How Does Lagrange Proceed?

Lagrange's idea that his theorem is correct is simple. It goes as follows: the essential argument is to show that alpha, beta, gamma, delta, ... in Euler's calculatory expression ‡ (p. 95) *can only* be taken by positive integers 1, 2, 3, ... However, this is self-evident, for (and that's the point!) in case of an exponent different from a positive integer the corresponding expression is *one-to-many*, whereas the given function, clearly, is one-to-one.—And this is all there is to it!

That is to say, Lagrange's argument is this: $x^{\frac{1}{2}} = \sqrt{x}$ has two value s ($\sqrt{4} = \pm 2$); $x^{\frac{1}{3}} = \sqrt[3]{x}$ has three values ($\sqrt[3]{1}$ has the value 1 and the two values $\frac{1}{2} \pm \frac{\sqrt{-3}}{2}$) etc. This is a real fact—if one is to accept the "complex" numbers.

The Fundamental Gap in Lagrange's Proof

Nevertheless, Lagrange's proof has a gap. This gap is of a fundamental nature and can be described as follows: Lagrange resorts to the argument that the "function" $x^{\frac{1}{2}}$ has *two* "values". However, only the following is true: there exist two "values" X to be discovered from the instruction $x^{\frac{1}{2}}$ ($= \sqrt{x}$), i.e. $X^2 = x$. (Example: we have $2^2 = 4$ as well as $(-2)^2 = 4$—and *consequently* two values for X.) But calculating is one thing—and mathematics another!

Lagrange did not set out to present an elementary *calculation,* but to *prove* a theorem! And within his proof he clearly *uses* the notion "value of a function"— e.g. by arguing that $x^{\frac{1}{3}}$ has *three* values. This argument is *wrong,* if one only considers "real" functions—as we do today in our first course at university. Three *real* numbers (i.e. those without the component $\sqrt{-1}$), which can be calculated following the instruction $x^{\frac{1}{3}} = \sqrt[3]{x}$ do not exist.

Well, you might object and say: Lagrange does not restrict himself to *real* functions—and, consequently, this objection is irrelevant.

The objection is only true in so far as Lagrange, indeed, did not write a theory of *real* functions, but (as we say today:) a theory of *complex* functions. Nevertheless, Crelle's argumentation is too weak, as it only deals with the given example (restriction to a *real* analysis) but not the essence. The essence remains:

Nowhere does Lagrange specify the concept "value of a function".

However, he implicitly *uses* the concept. This is definitely a deficiency in mathematical rigour.

You may say: what a tiresome thing, "value of a function"—isn't it obvious, what this means?

Counter-question: is it really so obvious? What is the "value of the function" for $\frac{1}{x}$ in case of $x = 0$? Or in case of $x = \infty$? That is to say: what is $\frac{1}{0}$, what is $\frac{1}{\infty}$? And moreover: what is the value of the function log 0, or tan $\frac{\pi}{2}$? Or of sin ∞? *This we do not know beforehand or through calculating,* these answers need a theory—and that means: we are to do some mathematics!

That is why we have to launch some criticism at Lagrange: nowhere do you declare what you take the notion "value of a function" to be! And it is unacceptable in mathematics to base a proof on an *undefined* notion!

Usually, Lagrange is criticized for something quite different. He is accused of making a technical mathematical error: that he has not taken into consideration that the representation § does not work *in particular cases.*

However, this criticism is completely unfounded. A first-rate mathematician such as Lagrange does not make such a technical mistake. If one reads his book *carefully,* one can see that he treated the said technical problem extensively and explained what he thought about it. Therefore, he knew about this problem—but this aspect does not touch his theorem and, in particular, it does not refute it. Because, just like with Euler, Lagrange's theorem is about "quantities", but not about particular "values". Unfortunately, today's mathematicians (as well as historians of mathematics) often do not follow this argumentation.

Literature

d'Alembert, J.-B.R. (1757). Grandeur. In *Encyclopédie, ou Dictionnaire Raisonné des Sciences, des Arts et des Métiers,* vol. 7, pp. 855. Paris: Briason, David, Le Breton, Durand.

d'Alembert, J.-B.R. (1761). Recherches sur les vibrations des cordes sonores. In *Opuscules mathématiques,* vol. I, pp. 1–64. Paris: David. https://gallica.bnf.fr/ark:/12148/bpt6k62394p.

d'Alembert, J.-B.R. (1771). Quantité. In *Encyclopédie, ou Dictionnaire Raisonné des Sciences, des Arts et des Métiers,* vol. 13, p. 655. Paris: Samuel Fauche.

d'Alembert, J.-B.R. (1780). Sur les fonctions discontinues. In *Opuscules mathématiques,* vol. VIII, pp. 302–308. Paris: David. https://www.bnf.fr/fr/collections_et_services/reproductions_document/a.repro_reutilisation_documents.html.

de Condorcet, M. (1771). Sur la détermination des fonctions arbitraires qui entrent dans les intégrales des équations aux différences partielles. *Histoire de l'Académie Royale des Sciences,* Année 1771, pp. 49–74.

Euclid. (1908). *The thirteen books of Euclid's elements.* In T. L. Heath, vol. I. Cambridge at the University Press.

Euler, L. (1828, 1829, 1830). *Leonhard Euler's vollständige Anleitung zur Integralrechnung,* 3 vols. Wien: Carl Gerold. German: Joseph Salomon.

Euler, L. (1885). *Einleitung in die Analysis des Unendlichen von Leonhard Euler.* Berlin: Springer. Reprint 1983. German: H. Maser. https://gdz.sub.uni-goettingen.de/no_cache/dms/load/toc/?IDDOC=264689.

Euler, L. (1763/68). E 342 *Institutiones calculi integralis, vol 1.* In F. Engel, & L. Schlesinger (Eds.), *Leonhardi Euleri Opera Omnia,* vol. 11 of series I. Leipzig, Berlin: Teubner, 1923. https://mdz-nbn-resolving.de/bsb00034024.

Grabiner, J. V. (1981). Changing attitudes toward mathematical rigor: Lagrange and analysis in the eighteenth and nineteenth centuries. In *Epistemological and social problems of the sciences in the early nineteenth century*, pp. 311–330. Dordrecht, Boston, London: D. Reidel.

Lagrange, J. L. (1813). *Théorie des Fonctions Analytiques*. In *Œuvres,* publiées par les soin de J.-A. Serret, vol. 9. Paris: Gauthier-Villars, 1881. Reprint: Olms, Hildesheim, 1973. https://books.google.de/books?id=XGQSAAAAIAAJ& pg=PA4&ots=ZgppLb1box&dq=lagrange+22theorie+des+fonctions+analytiques22& lr=, https://gallica.bnf.fr/ark:/12148/bpt6k86263h/f1.table. Text version (extracts): https://gallica.bnf.fr/ark:/12148/bpt6k88736g. German: Lagrange 1823.

Lagrange, J. L. (1823). *Theorie der analytischen Functionen.* J. L. Lagrange's mathematische Werke, vol. 1. Berlin: G. Reimer. German: A. L. Crelle.

Chapter 9
Bolzano: The Republican Revolutionary of Analysis

The Situation

From the Academies to the University

The Algebraic Analysis of the eighteenth century was created by mathematicians at academies: Euler, d'Alembert, and Lagrange were prominent as members of scientific academies, but not as university professors. This changed with the growing industrialization of the economic life and with increasing mechanization at the turn of the century. Society needed more engineers of all kinds, and, as modern technology relies on mathematics, higher education for engineers became necessary.

What effect this had on the centre of mathematical development at that time, Paris, will be shown in the next chapter. First we have to take a detour to Prague and to the province of Bohemia.

Bolzano: The Public Enemy

Emperor Francis had ordinal number I as Emperor of Austria and ordinal number II as the last Holy Roman emperor. By a ruling from 24th December 1819 and notification on 20th January 1820, he dismissed the theologian, philosopher and mathematician Bernard Bolzano (1781–1848) from his chair of philosophical theology at the University of Prague. Bolzano was kept under surveillance and banned from publishing. Aid of the local gentry, among them functionaries of the Church, prevented the worst and allowed Bolzano to pursue his studies in the absence of his university position.

© The Author(s), under exclusive license to Springer Nature Switzerland AG 2022   101
D. D. Spalt, *A Brief History of Analysis*,
https://doi.org/10.1007/978-3-031-00650-0_9

Bolzano's thoughts were considered too dangerous. The Austrian Emperor wanted to prevent an overthrow of the social order, as had happened in France, at any price. Bolzano's sermons as a priest were thought to be socially revolutionary and the Emperor's confessor suspected subversion. Had not a priest and Prague University professor Jan Hus (c1370–1415) inspired ordinary people through his sermons and made himself an enemy of the authorities some 400 years earlier?

That Bolzano was a first-rate revolutionary thinker is also evident from those mathematical manuscripts, which he composed in seclusion from the scientific world and which were finally printed in the late twentieth century. We shall come back to that.

A New Meaning of Convergence

Today, "convergence" is a principal concept of analysis. And as we know already, this had not always been the case.

Euler: A Reminder

We have already seen Euler's understanding of "convergence" (p. 84): a series $a_1 + a_2 + a_3 + \ldots$ "converges" (in Euler's sense), if the a_i continually decrease and get arbitrary small, as e.g. $1 + \frac{1}{2} + \frac{1}{3} + \frac{1}{4} + \ldots$ (p. 85). However, in Euler's analysis, "convergence" did not play any part.

Today

Today things are different—at least in scientific textbooks. There, the notion of "convergence" (of a series) is often introduced quite early, even before the notion of "continuity" and even before the techniques of differentiation and integration.

In textbooks for engineers, things may be different, or in cases where the author aims to present analysis as a *very first encounter* with mathematics and not as a *prelude* to it.

An author of the latter kind is Michael Spivak. His textbook *Calculus* published in 1967 comprises 29 chapters and some 500 pages. He presents the concept of "convergence" as late as in chapter 21, on page 373, only after the introduction and treatment of "continuity", "derivative", and "integral". It appears that modern analysis can be constructed without using the concept "convergence".

The Convergence of Sequences: Two Notions

We saw: in the seventeenth and eighteenth century, the mathematicians worked with *series:* $a_1 + a_2 + a_3 + \dots$ Today we are more used to begin with *sequences*—these are the *separate* summands of a series in their succession: $a_1, a_2, a_3, \dots$; they are not added.

Today we define the concept "convergence" of such a sequence in two steps (we read Spivak):

1. "Convergence towards a limit":

A sequence $a_1, a_2, a_3, \dots$ "converges to" l (in symbols: $\lim_{n \to \infty} a_n = l$) if for every $\varepsilon > 0$ there is a natural number N such that, for all natural numbers n,

$$\text{if } n > N, \text{ then } |a_n - l| < \varepsilon.$$

l is called the "limit" of the sequence.

Without worrying about all the details of this definition, one soon gets the impression: this definition is very similar to that which Euler has given in *his* conceptual treatment of series. He had demanded that the a_n decrease continually and finally vanish. Speaking with Spivak: Euler demands that the "sequence" of the a_n *converges to the limit* 0: $|a_n| = |a_n - 0| < \varepsilon$—written in today's notation: $\lim_{n \to \infty} a_n = 0$.

2. "Convergence" without a "limit":

A sequence $a_1, a_2, a_3, \dots$ "converges" if for every $\varepsilon > 0$ there is a natural number N such that, for all m and n,

$$\text{if } m, n > N, \text{ then } |a_n - a_m| < \varepsilon.$$

(Today such sequences are usually called "Cauchy sequences"—although Bolzano was earlier, as we shall see soon.)

This gets really complicated! How should we understand this?—We shall postpone the answer until we have looked at Bolzano's approach and carry on.

It is not allowed to define a concept twice! That is why we have to disentangle these two steps. To begin with:

1. *Each* of the two formulations in the boxes is a *possible* definition of our modern concept "convergence".
2. The first formulation is more specific and the second more general. (Because the first formulation *has* the "limit", whereas in the second case one must *prove* that such a limit l exists at all!)

3. Therefore, let us *prove* that both formulations are equivalent *for Bolzano*!

   (a) *The first formulation implies the second:* we suppose that $|a_n - l|$ can be made less than any given (positive) value, starting from a natural number N. Consequently, $|a_n - l|$ as well as $|a_m - l|$ can be made $< \frac{\varepsilon}{2}$, if only $n > N$ and $m > N$. The calculation is simple if we resort to the "triangle inequality" ($|x \pm y| \leq |x| + |y|$). Because for $n, m > N$, we have

   $$|a_n - a_m| = |a_n - l + l - a_m| \leq |a_n - l| + |l - a_m| < \frac{\varepsilon}{2} + \frac{\varepsilon}{2} = \varepsilon,$$

   and that was to be proved.

   (b) Now the *reverse direction*. We assume that $|a_n - a_m|$ can be made arbitrarily small *from some natural number N onwards* (formally, for $n > N$ and $m > N$). Where do we get l from without stealing it?—Bolzano has an idea. He says that l can be "determined as precisely as one wishes". And he thinks this to be sufficient. Let us decide that for ourselves!

   (i) It is, *indeed,* possible to determine l with any degree of accuracy. We only have to determine a suitable ε. For an accuracy of a tenth, we choose $\varepsilon = \frac{1}{10}$. Then it follows from our assumption that there exists a natural number N such that for $n > N$ and $m > N$ we have $|a_n - a_m| < \frac{1}{10}$. So we simply choose $l = a_n$ and we obtain $|l - a_m| < \frac{1}{10}$, as required.

   (ii) Without a *conceptual construction* of "number", no better argument is known so far.

   (iii) As we shall see later, not until the year 1872 (or maybe 1849 or 1838) was such a *conceptual construction* of "number" forthcoming in mathematics.

   (iv) That is why Bolzano obtained the best possible result for his time. In short, the reverse direction is also proved, and that by Bolzano himself in the year 1817; we shall come to this in the next but one section!

The Convergence of Series Today

In his chapter 22, Spivak defines the concept "convergence" also for *series*. Spivak is very precise and up to date. Therefore, he does not say in his definition that the "series" $a_1 + a_2 + a_3 + \dots$ "converges", but instead the "sequence" $a_1, a_2, a_3, \dots$ is "summable" (the meaning is the same):

> The sequence $a_1, a_2, a_3, \ldots$ is "summable" if the sequence $s_1, s_2, s_3, \ldots$ converges, where
>
> $$s_n = a_1 + \cdots + a_n .$$
>
> In this case, $\lim_{n \to \infty} s_n$ is denoted by
>
> $$\sum_{n=1}^{\infty} a_n \quad (\text{or, less formally,} \quad a_1 + a_2 + a_3 + \cdots)$$
>
> and is called the "sum" of the sequence $a_1, a_2, a_3, \ldots$

Is this completely clear? Maybe not?

You might be tempted to think that the "convergence" of the series $a_1 + a_2 + a_3 + \cdots$ is about $(a_1 + a_2 + a_3 + \cdots + a_n)$ for $n \to \infty$—but this is not the case! To understand the real meaning of " $\lim_{n \to \infty} s_n$ ", one has to examine the definition of the second step above more deeply and this becomes complicated!

Instead of worrying about this any longer, let us now turn to Bolzano, because he has defined *exactly the same* notion ("convergence" of a "series", the "summable" of Spivak), but in a linguistically much more ingenious way which can be understood much more easily.

The Convergence of Series by Bolzano

An almost unnoticed mathematical treatise from Bolzano appeared in 1817 in Bohemia (far from Paris), for which Bolzano is still praised today. This praise is completely justified, as we shall soon see.

Bolzano did not give the concept any name (such as "convergence"). He directed the reader's attention to this property by calling it "particularly remarkable":

> Particularly remarkable among such series is the class of those series with the property that the *change* in value (*increase* or *decrease*), *however far the continuation* of its terms is taken, always remains *smaller* than a certain quantity, which itself can be taken *as small as we please,* provided the series has been continued far enough beforehand.

A lengthy and long-winded sentence (*emphases* are not at all added). Unfortunately, this is characteristic of Bolzano. Nevertheless, this is a fascinating, clearly made definition! Let us analyze this sentence in more detail.

1. The crucial passage is "however far the continuation of its terms is taken". Bolzano writes it as a formula in the following way:

$$a_n + a_{n+1} + a_{n+2} + \ldots + a_{n+r} . \tag{¶}$$

Spivak notes for this (caution: the convergence of series is about the estimating of the magnitude of the sums s_n but only *indirectly* about the a_n!):

$$|s_{n+r} - s_{n-1}| \, .$$

Although it says the same, this is much less understandable. Really! (Is this true?)

2. "However far the continuation is taken" means: r is allowed to get arbitrarily large; there are allowed arbitrarily many, *very many* summands (but of course, only *finitely* many).

3. By the "however far the continuation is taken" (that is the sum $\P$), Bolzano demands it must "always remain smaller than a certain quantity, which itself can be taken *as small as we please*". Today we formulate the latter as seen already in Spivak: "for all $\varepsilon > 0$ we have $| \ldots | < \varepsilon$", and the three dots stand in for almost the entire sum $\P$.
   Today the pompous formulation "for all ..." is called the "universal quantifier". Bolzano gives a completely different formulation, namely this: "... a certain quantity [i.e. ε], which itself can be taken *as small as we please*". This is much smoother than the brutish "all": "a certain ..., which itself can be taken ...". (That quantities do not have *negative* values—Spivak's "> 0"—was in the mathematics of the time of Bolzano a matter of course and, consequently, not written down.)

4. The final touch: the phrase "can be taken" implies a *basic assumption,* namely, the series "has been continued far enough" *beforehand.* What does Bolzano consider here?—The n: with his formulation "has been continued far enough" he means: *in order that* the sum $a_n + a_{n+1} + a_{n+2} + \cdots + a_{n+r}$ remains *always*—for each r—smaller than the chosen ε, the n has to be taken *large enough*.
   This condition is also formulated with the help of a *quantifier* today, namely with the other one, the "existential quantifier": "there exists a number N such that ...". Again, Bolzano formulates this more simply: "the n must be chosen great enough". The content is the same: what Spivak gives in a normed, pompous formulation, Bolzano puts in simpler prose, but not in colloquial language. Unfortunately, such prosaic language is impossible in case of intricate notions such as "convergence". Mathematics does not come without effort!

5. Recapitulation: if we now reread Bolzano's sentence about the "particularly remarkable" class of series, we would be able to enjoy it, wouldn't we? In this *single* sentence, Bolzano embraces everything for which Spivak needs three stages of a definition (the last one on p. 105). Additionally, *Bolzano presents the most general definition of "convergence"*. He does this in a single sentence of five (in the German original, seven and a half) lines. Of course, Bolzano's sentence is somewhat *challenging,* in the vocabulary as well as in the syntax. Nobody understands it at once. One has to read it repeatedly. But it can be understood without any previous knowledge (especially of logic). This sentence

is not technical at all and it lacks modern quantifiers. Therefore it is open to the laymen and the laywomen. Let us note:

> The modern concept "convergence", in its most general version, was first formulated by Bolzano in the year 1817. (See the previous box as well as the proof on pp. 104f.)

In science it is common to take the year of publication as the year the text was created because in many cases (not in this one!) it is hard to detect in which year the manuscript was *really* finished. The publishing date is usually undisputed.

The Remaining Deficiency

Although Bolzano argued very subtly and conceived his notions very carefully, he capitulated at the same point as once Leibniz did (see p. 31): just like his renowned predecessor, he also, one and a half centuries later, lacked a concept of number which served his needs. As stated before, to *calculate* a number "as precisely as one wishes" is not the same as to *have* a number, to *present* it! To carry out its proofs, mathematics needs a *proper concept* of number (which itself can be constructed in a proof); otherwise it cannot do so. In 1817, Bolzano had none. (Only half a generation later, he worked on this project, but this remained hidden in his manuscripts until the last quarter of the twentieth century.)

Continuity with a New Meaning

Convergence Works with Discrete Objects

The notion "convergence" is made for *discrete* objects.

In the nineteenth century, those discrete objects were the *summands* of the infinite series. The series $a_1 + a_2 + a_3 + \ldots$ consists of the *discrete* objects: $a_1, a_2, a_3, \ldots$

Towards the end of the nineteenth century, these discrete objects *as they stand* were taken as genuine mathematical entities in their own right and they got a name, which was never used before: "sequence". (This way a new mathematical object was created.) The "sequence" is $a_1, a_2, a_3, \ldots$, and a new notation was invented: "$(a_n)_n$", or sometimes in full length: "$(a_n)_{n \in \mathbb{N}}$", whereby "$\mathbb{N}$" denotes the set of natural numbers. Some authors use braces instead of round brackets and sometimes the running index is omitted, and then the sequence is just "$\{a_n\}$", more on this in Chap. 13.

Continuity Is Analogous to Convergence

Anybody who understands the concept of "convergence" will recognize at once the meaning of "continuity".

> The transfer of the modern concept of "convergence" into the continuum creates today's concept of "continuity".

The notion "continuity" is more easily visualized. Therefore, if you have difficulties with the notion "convergence", you should study the notion "continuity" and return thereafter to "convergence".

The German language has two expressions, whereas the English as well as the French only has one: the German calls the *new technical term* "stetig" as distinguished from "kontinuierlich", whereas English and French only have "continuous" and "continue", respectively.

Continuity of Functions in Bolzano

Bolzano defines continuity to be the following property: a function is called "continuous" at the value x_0, if

> the difference $f(x_0 + \omega) - f(x_0)$ can be made smaller than any given quantity, provided ω can be taken as small as we please.

Bolzano takes ω, the last letter of the Greek alphabet, as a variable which can be "taken as small as we please". Without stating it explicitly, Bolzano always assumes a positive ω: $\omega > 0$.

Let us construe the function $f(x)$ not only as a *calculatory expression* but also as a drawn *curve* in the coordinate system. If so, Bolzano's definition says: at the value x_0, the *curve* does *not have a sudden change*. The more the x-values approach the value x_0 (i.e. the *smaller* ω gets), the smaller is the difference of the respective values of the function $f(x + \omega)$ from the value $f(x_0)$; expressed formally,

$$\text{if} \quad |\omega| < \delta \quad \text{we have} \quad |f(x_0 + \omega) - f(x_0)| < \varepsilon .$$

Bolzano says: "The difference [namely: $|f(x_0 + \omega) - f(x_0)|$] can be made smaller than any given quantity, *provided* ω can be taken as small as we please.". It is a logical challenge to grasp this. It says that the δ is to be defined *after* ε is given, at first the ε and then the δ. Can we accept this?

As it turns out, at this point Bolzano is a little bit sloppy! He *denotes* the "value" of the variable x in the same way as the "variable" itself, plainly by "x". The notation "x_0" for a particular "value" of the variable x is my doing, it is not in Bolzano. (I adopted it from Cauchy, as will be seen in the following chapter, p. 123.) As far as I know, nobody was ever disturbed by Bolzano's carelessness; everybody understood him correctly. But I want to proceed here as pedantically as it is (meaningfully) possible.

The Little Difference Between Then and Now

Bolzano's definition of "continuity" is neat, clear, and still valid today. Nevertheless, for his contemporaries it meant something other than for us today, because they unspokenly assumed:

> A quantity does not have negative values.

At the beginning of the nineteenth century, negative numbers were still not accepted as true numbers in mathematics. (See also p. 27.)

In his definition, Bolzano unequivocally writes "$x_0 + \omega$" instead of "$x_0 \pm \omega$". (And of course, with "$+$"Bolzano also *means* +, not "+ or −".) The implication is as follows: Bolzano studies only the *right side* of x_0, that is to say only, those values that are *greater* than x_0. The *smaller* ones (i.e. $x_0 - \omega$) he ignores. Consequently, Bolzano only defines *one-sided* continuity.

The scrupulous thinker Bolzano was clearly aware of this. In his later work *Theory of Functions*, he distinguishes between "positive increments'" and "negative increments", and similarly he differentiates between continuity "in the positive direction" and "in the negative direction". If both apply, he uses the words "plainly continuous". (However, in his formulae, he only uses the plus sign.)

Such a strange thing as alternating signs within the terms of the sequence, as for instance in the sequence $1, -\frac{1}{2}, \frac{1}{3}, -\frac{1}{4}, \frac{1}{5}, -\frac{1}{6}, \ldots$, could not be imagined even by somebody as revolutionary as Bolzano at that time. This was not due to a lack of radicalism in Bolzano's thought, not at all! The explanation is this: according to Bolzano, the *continuous* variable is, of course, only capable of *continuous* changes! Naturally, Bolzano thinks of the quantity ω, denoting an "increment" of the continuous variable x, as being continuous, too. In all probability, his contemporaries did the same. (The mathematical object "sequence" did not exist at the beginning of the nineteenth century, and it was not invented yet.)

The Differences from Euler's Continuity

Euler called a *"curved line"* "continuous" if it could be described by a calculatory expression (see p. 81). Bolzano calls a *"function"* "continuous" if its geometric drawing shows no sudden change.

These two concepts of continuity differ in two respects:

1. They relate to different *objects:* Euler deals with "curves" and Bolzano with "functions".

2. They denote completely different *properties:* for Euler it is the property that a curve can be described *as a whole* by a single calculatory expression, and for Bolzano it is the property of a function to change its values *near to another value* only very gradually.

That is why Euler's concept of continuity is sometimes called "global" (it relates to the *whole* curve), while Bolzano's concept is called "local" (because it only operates in the neighbourhood of a *single* value x_0). However, a helpful comparison of both concepts is difficult, as they deal with totally different objects (curve *resp.* function). It gets even worse if one tries to include d'Alembert's idea about "not continuous": p. 93, item a.

Sometimes one is a bit less rigorous and *identifies* a "function" with its *drawing* in the coordinate system; this lessens the differences of the two notions of continuity. But mathematics should have no place for inaccuracies!—however, see p. 125...!

Euler's concept of continuity, together with his "Algebraic Analysis", perished a long time ago. The concept of continuity has survived, as it was created for the first time by Bolzano. This is not the merit of Bolzano. On the contrary, this shows: *in a certain sense* our analysis today is the same as Bolzano's—but completely different from Euler's.

Continuity and the Continuous

The new concept "continuity" is conceived for dealing with the *continuous*. It transfers the notion "convergence" from the *discrete* to the *continuous*. This we shall point out shortly.

"Continuity" demands for the function $f(x)$:

$$\text{for} \quad |\omega| < \delta \quad \text{we have} \quad |f(x_0 + \omega) - f(x_0)| < \varepsilon.$$

Now we concentrate on two analogies:

1. Instead of the *continuous* object "variable", i.e. x, we take the *discrete* object "natural number"; the individual may be named n.
2. Instead of the *continuous* object "function", i.e. instead of $f(x)$, we take the *discrete* object "sequence", i.e. $(a_n)_n$.

Consequently, the fixed value of the function $f(x_0)$ corresponds to the limit l, and the continuously varying values of the function $f(x_0 + \omega)$ correspond to the discrete terms a_n of the sequence.

$$\text{Then} \quad |f(x_0 + \omega) - f(x_0)| < \varepsilon$$

$$\text{changes to} \quad |a_n - l| \quad < \varepsilon.$$

Finally, the condition to determine a sufficiently small δ to a given ε such that $|\omega| < \delta$ is guaranteed for the first inequality translates to the condition to determine a number N great enough such that for $n > N$ the second inequality is valid. —Done!

Bolzano's Revolutionary Concept of Function

Even after his relegation from university, Bolzano had the possibility of continuing his studies. (Only a few succeed in this.) Some 17 years after the publication of the treatise for which he is still praised today, his mathematical manuscripts were completed sufficiently to provide a foundation for his work on a textbook of analysis. The title of this manuscript reads simply: *Theory of Functions*.

Bolzano had already laid those foundations in earlier papers. Here the mature thinker Bolzano reached a notion of function, which is quite removed from everything that his contemporaries were able to think about.

Because this innovative work stayed unknown until the twentieth century, it could not be of any influence within the world of mathematics. However, it is spectacular to such a degree that we will give a brief comment on it below.

Bolzano's Definition of the Concept of Function

As the reader might already expect, Bolzano expresses this definition also by a single (somewhat lengthy) sentence:

> The variable quantity **W** is a *function* of one or several variable quantities **X, Y, Z**, if there exist certain propositions of the form: the quantity **W** has the properties **w, w₁, w₂**, which can be derived from certain propositions of the form: the quantity **X** has the properties ξ, ξ', ξ''—, the quantity **Y** has the properties η, η', η'', etc.

If we translate this long-winded language of Bolzano's *Biedermeier* period into today's words, it might read as follows:

> A variable quantity w is called a "function" of the variable quantities x, y, z, if the properties of w are determined by the properties of x, y, z via a law which depends on certain values.

Or, more succinctly:

> A quantity w is called a "function" if its values are determined by a law which depends on the values of other variables.

This is the *most general* notion of function (in analysis) possible and this even by today's conceptions and standards. It is crucial that *the "calculatory expression"* (as in Euler) *is no longer mentioned!* Only a "law" is required, a "dependency from certain values" (Bolzano uses the word "derivable").

Bolzano's Examples of Functions

Bolzano not only jots down this definition, he actually means it in this generality. One can see that already in his *Theory of Functions* by studying his very first example of a function:

> If **W** denotes the prize that shooters are rewarded with for skill at target practice if we decide that a shot in the bullseye should get 100 *Reichsthaler*, a shot which is a distance $= x$ inches from the bullseye, where x is not more than 2, gets $100 - 25x$ *Reichsthaler*, and a shot with distance from the bullseye > 2 and < 5 gets $58 - 2x^2$ *Reichsthaler*, etc.

Judgement

It is not the (very simple) *calculatory expressions* used by Bolzano which are important, but the *arbitrary piecewise* manner of the definition of the winnings.

For Bolzano's contemporaries (like Cauchy—we shall meet him in the next chapter), such a way of specifying the mathematical concept of "function" was *completely unthinkable* and downright out of place. Analysis should deal with such objects? "Anything but that!" was the opinion of Bolzano's contemporaries.

Bolzano was decades ahead of his time with this idea. Riemann's definition from the year 1854 *clarified* one aspect of Bolzano's definition but actually did not add anything new to it. However, Riemann's formulation was printed and became well-known and marks the state of affairs to this day. (We shall come to that on p. 155.)

Mathematical Consequences of Bolzano's Notion of Function

Another demonstration that Bolzano really *meant* his notion of function to be so radical can be found in a casual half-sentence in his *Theory of Functions* (there Bolzano uses even the formulation "dependency"):

> As it is permitted to think the law of dependency of a number from another as we wish ...

Then Bolzano gets creative. At first he constructs a (simple) function, which is not "continuous" for *any* chosen interval of x, be it as small as one wishes.

In contrast to this, his contemporaries were convinced that a function is usually *nearly everywhere* "continuous"—with the exception of some particular values. Bolzano constructed a radical counter-example to this way of thinking.

But this is still not enough! Bolzano constructs other functions with completely unexpected properties—e.g. one which "in between the limits of $x = 0$ and $x = 1$ infinitely often increases or decreases".

Finally, Bolzano constructs (by slightly overstretching his notion of function) a function, which for any interval of x, however small, is neither *only* increasing nor *only* decreasing during this interval. (Today we say: which is not "monotonous" in any interval, however small it is taken.)

This function invented by Bolzano is a true mathematical monster. It is continuous, but it is in no interval monotonous. Even worse, it cannot be differentiated *at any value.*—Contemporaries and successors of Bolzano were thoroughly convinced that a function could be differentiated (nearly) *everywhere,* some of them even as late as in 1874: see p. 171. Bolzano's thinking was far more radical—more like ours today.

Literature

Bolzano, B. (1816). *Der binomische Lehrsatz, und als Folgerung aus ihm der polynomische, und die Reihen, die zur Berechnung der Logarithmen und der Exponentialgrößen dienen, genauer als bisher erwiesen.* Prag: C. W. Enderssche Buchhandlung. http://dml.cz/handle/10338.dmlcz/ 400346. English in: Russ 2004, pp. 154–248.

Bolzano, B. (1817). *Rein analytischer Beweis des Lehrsatzes, daß zwischen je zwey Werthen, die ein entgegengesetztes Resultat gewähren, wenigstens eine reelle Wurzel der Gleichung liege.* In P. E. B. Jourdain (Ed.), Leipzig, 1905. http://dml.cz/manakin/handle/10338.dmlcz/400352? show=full. English in: Russ 2004, pp. 249–277.

Bolzano, B. (1830–1834). *Reine Zahlenlehre.* In J. Berg (Ed.), *Bernard-Bolzano-Gesamtausgabe,* vol. II A.8. Stuttgart–Bad Cannstatt: Friedrich Frommann Verlag (Günther Holzboog), 1976.

Bolzano, B. (1830–1835). *Erste Begriffe der allgemeinen Größenlehre.* In J. Berg (Ed.), *Einleitung zur Größenlehre und Erste Begriffe der allgemeinen Größenlehre. Bernard-Bolzano-Gesamtausgabe,* vol. II A.7, pp. 217–285. Stuttgart-Bad Cannstatt: Friedrich Frommann (Günther Holzboog), 1975.

Bolzano, B. (1834–1842a). *Einführung in die Funktionenlehre.* In K. Rychlík (Ed.), *Bernard Bolzano's Schriften,* vol. 1. Prag: Königl. Böhmische Gesellschaft der Wissenschaften, 1930. http://dml.cz/handle/10338.dmlcz/400333.

Bolzano, B. (1834–1842b). *Functionenlehre.* In: B. van Rootselaar (Ed.), *Bernard-Bolzano-Gesamtausgabe,* vol. II A.10/1. Stuttgart-Bad Cannstatt: Friedrich Frommann (Günther Holzboog), 2000. English: Russ 2004, pp. 429–589.

König, G. (Ed.) (1990). *Konzepte des mathematisch Unendlichen im 19. Jahrhundert.* Göttingen: Vandenhoeck & Ruprecht.

Russ, S. (2004). *The mathematical works of Bernard Bolzano.* Oxford: Oxford University Press.

Spalt, D. D. (1990). *Die Unendlichkeiten bei Bernard Bolzano.* König 1990, pp. 189–218.

Spivak, M. (1967). *Calculus.* Amsterdam: W. A. Benjamin, Inc.

Chapter 10
Cauchy: The Bourgeois Revolutionary as Activist of the Restoration

Cauchy: The Atipode to Bolzano

Augustin-Louis Cauchy (1789–1857) was taught mathematics between the ages of 15 and 18 at the École Polytechnique in Paris, by first-rate teachers of his time. At the age of 26, he himself started teaching at this university. In the meantime, he worked as an engineer, for instance, in Paris on the construction of canals and an aqueduct and in Cherbourg on the construction of the new harbour.

This is an extraordinary biography: a singularly gifted mathematician works in his earliest years for several years in the material construction of the world *through* mathematics until he starts his creative mathematical career. This had unique consequences on the mathematical disposition of the man Cauchy as well as on the transformation of mathematics as a science through Cauchy.

For the sake of completeness, we may add: at the age of 26, after the downfall of Napoleon and the beginning of the Restoration, Cauchy (together with a physicist) was appointed by King Louis XVIII to the Académie des Sciences—this after the exclusion of 62-year-old Lazare Carnot and 69-year-old Gaspard Monge for political reasons.

The Heart of Cauchy's Revolution of Analysis

As a mathematician, Cauchy was influential not so much as a member of the Academy but as a university teacher. Accompanying his lectures in 1821, he published a textbook on Analysis (*Cours d'Analyse de l'École Royale Polytechnique*), in 1823 a textbook on differential calculus, and in 1829 one on integral calculus. (Cauchy counts as one of the most productive scientists. In addition, he wrote plenty of treatises; today's edition of his *Collected Works* comprises 27 large sized, thick volumes.)

© The Author(s), under exclusive license to Springer Nature Switzerland AG 2022 115
D. D. Spalt, *A Brief History of Analysis*,
https://doi.org/10.1007/978-3-031-00650-0_10

In the introduction of his textbook on Analysis, he drew conclusions from his experience as an engineer who did practical work by writing the following two sentences (the *emphases* are added):

> We must also observe that [the arguments drawn from the generality of algebra] tend to grant a limitless scope to algebraic formulas, whereas, in reality, most of these formulas are valid only under certain conditions or *for certain values* of the quantities involved. In determining these conditions and *these values* and in establishing precisely the meaning of the notation that I will be using, I will make all uncertainty disappear, so that the different formulas present nothing but relations among real quantities, relations which will always be easy to verify *by substituting numbers for the quantities themselves.*

> There had never been a more radical declaration of war on traditional mathematics in modern times.

The radicalism of this challenge was so comprehensive that nobody was aware of it or even took it seriously, neither Cauchy's contemporaries nor later historians of mathematics.

This is true even though Cauchy said everything in these two sentences. For what do they say?

The *first sentence* says that the formulae of the Algebraic Analysis of Lagrange and Euler are completely general (they "tend to grant a limitless scope")—while practical mathematics (the "reality") shows that these formulae are valid *only* "for certain values". In short, the mathematics from the Academies is vague, aloof, and removed from reality—its application to practical problems requires the precise determination of their validity.

The *second* sentence says two things: (i) the meanings of the mathematical *terms* must be clear and unequivocal and (ii) through the *assignment of the* VALUES *in the expressions*, we obtain crystal-clear numbers which are suitable *in practice*. And the *numbers* are such "values" by all means.

In the years he was working with others as an engineer, Cauchy *experienced* at first hand:

> The analysis only becomes *useful and effective* when its general *formulae* are adapted in such a way that they result in *numerical values.*

What was Cauchy to do when he was appointed to teach Analysis at the École Polytechnique to future engineers? He taught them to adjust the general formulae by *substituting numbers* in order to make them *useful for the practice.*

Mathematical View of Cauchy's Revolution of Analysis

Euler and Lagrange founded Algebraic Analysis on a most *general* concept of "variable". By wanting to go beyond this, Cauchy, *as a matter of principle,* declared that he wished to *materialize* the "variables" by assigning "values" to them. However, this amounts to nothing else than that:

> Cauchy wished to *add* to the original foundational concept of Algebraic Analysis "variable" a second and new one: the foundational concept "value".

Cauchy's *motive* for this innovation was a *non*-theoretical one: his practical experience in applying the formulae of Algebraic Analysis.

However, the *consequences* of Cauchy's move are deeply theoretical. Even the layman will understand that:

> A change in the foundational concepts of a theory *necessarily* changes the theory itself.

The reason is that "theory" in mathematics involves proofs. Yet the foundational concepts are the starting points of those proofs. Subsequently, it follows that new foundational concepts must change the proofs!

In other words, by announcing, *as a matter of principle,* the making of general formulae of the "Algebraic Analysis" fit for praxis by introducing values into the expressions, Cauchy complements the foundation of the theory (the "variable") by concretizing it to take specific "values".

Such a foundational reconstruction of a theory is nothing other than a *conceptual revolution.* The use of the term "revolution" is not just empty talk or a way of attracting attention; it shows the significance of Cauchy's innovative idea.

Science needs clear concepts. A *clearly determined* fact needs a special, unique name.

This also holds for the science history of mathematics. The analysis of Euler and Lagrange has been called "Algebraic Analysis"—we propose the name "Calculus of Expressions" (see p. 94).

Cauchy abolishes this kind of analysis and establishes a new theory instead. He states this *explicitly* in the *Introduction* of his first textbook of analysis. Therefore, Cauchy's analysis clearly needs a new name of its own!

Fascinatingly, history of mathematics has not come up with such a name until today: the analysis as introduced by Cauchy has not been christened yet. Therefore, we are free to do so. My proposal is as follows: let us call it "Calculus of Values" or **"Analysis of Values",** a name that draws attention to the structural change in analysis. Our result is:

> Cauchy transformed the "Calculus of Expressions" (or "Algebraic Analysis") into the "Calculus of Values" (or "Analysis of Values").

Cauchy's Concept of Variable Is Determined by "values"

Cauchy defines "variable" this way:

We call a quantity *variable* if it is assumed to take on many different values.

This is not alarming. But it clearly differs from Euler's declaration (p. 71). Euler demanded that "all definite values whatsoever" should be taken by the variable. Thus, since Cauchy a "function" may be defined only within a finite interval. We now operate within a "Calculus of Values" and no longer within a "Calculus of Expressions".

The graduate of mathematics realizes that Fourier Analysis is only possible since Cauchy—i.e. the method of finding a suitable calculatory expression in the form of an infinite trigonometrical series for an arbitrarily given function. The reason is that the construction of this series requires the calculation of *definite integrals* and this is only possible if the domains of the respective functions are *finite* intervals.

Euler, the indefatigable calculating mind nevertheless managed to construct the aforementioned "Fourier Integrals". However, as Euler was unable to conceive of a "function" with a finite interval as its domain, he was unable to recognize the wider scope of his calculation. This interpretation goes back to Jean Baptiste Joseph de Fourier (1768–1830) who wrote it down in the first quarter of the nineteenth century and who subsequently lent his name to this technique.

What cannot be seen in this definition of "variable" is its actual *usage* by Cauchy. At first sight, Cauchy's *designation* of the "variable" seems to be *hardly* different from that of Euler—apart from the already noted (and very *important!*) aspect that Cauchy no longer demands the variable to take "all definite values whatsoever". But beyond that?

The difference cannot easily be seen from Cauchy's *definition* of the notion of variable. The meaning of "value" introduced by Cauchy has further importance and lends a greater degree of precision than it had for Euler or Lagrange and their "Calculus of Expressions".

Cauchy Derives "number" from "quantity"

"Quantity"

Modern analysis is founded on the concept of "number". Yet, Cauchy could not build on this concept because he did not have a suitable notion of "number". He knew that the decimals are suitable and useful for *practical calculations.* Through the lack of an alternative idea, Cauchy stuck to tradition and relied, as did his predecessors, on the notion of "quantity". (For the proofs, mathematicians need *concepts.*)

Cauchy starts, in a similar way to his teacher Sylvestre François Lacroix (1765–1843), with "magnitudes" and focuses on their *changes,* more precisely on their *increases* or *decreases.* These changes of the "magnitudes" he calls by the traditional name *"quantity".* (The word "Zahlgröße" was coined in the German language for this type of quantity at that time.) We summarize:

> The "quantity" is the increase or the decrease of a "magnitude". The "opposite" quantity is the decrease or the increase which the second quantity undergoes to reach the first.

That means that "quantity" is a *change,* which is a notion that comprises *motion.* This is clever because the *concept* itself is fixed, motionless. It only *comprises* motion (and we already know from p. 27 that according to the standards of classical mathematics motion is not accepted in mathematics).

Let us look at the philosophical aspect: Cauchy's definition, just like Euler's (see p. 70), does not say *what* "magnitude" essentially ought to be.

"Number"

Cauchy bases the concept "quantity" on the notion "magnitude". In the same way, he deals with the concept "number". He writes:

> The measure of the second magnitude, if compared to the first, is a *number* which is represented by the *geometrical ratio* of one to the other.

If, e.g., the two measures are "6 hours" and "2 hours", their geometrical ratio is the "number" 3:

$$\frac{6 \text{ hours}}{2 \text{ hours}} = \frac{6}{2} = 3 \, .$$

This does not seem convincing: doesn't the measure "6 hours" already contain the number 6?—Strictly speaking, this is not the case! The ancient inhabitants of Mesopotamia arranged their civic and economic activities as well as their fortifications, canal planning, and accounting (the documents in clay have survived until today) *without knowledge of the abstract concept of "number"*. For anything and everything, they had their own system of measure; these systems were totally different from each other. The signs for "6 days" and for "6 goats" were, for instance, completely different.

Here is not the place to go into further detail, but those thousand years of early civilization show clearly: the operation with *measures* does not *necessarily* require the knowledge of the abstract object of "number". *Measuring* and (abstract) *counting* are completely different actions. Accounting does not need numbers! Historically, measuring preceded counting by specialists (with numbers) by thousands of years.

Cauchy dealt with the concept of number only to a certain degree. (He asked, for example, what about 0? Following the existing concept, zero is no number!) We are content with this answer: this way, Cauchy was at least able to give a clear concept of fractions (or rational numbers).

However, for doing analysis, Cauchy needs more, at least the "irrational" numbers. How did Cauchy think about them?—For this, we need some preliminaries.

> The central conceptual tool for the construction of analysis introduced by Cauchy is the concept of **"limit"**.

The mathematical abbreviation is "lim", standing for the Latin "limes" and the French "limes" and "limite".

The Basic Definition of "limit"

The object is not new. We have already found it in Leibniz, with regard to his convergence criterion (as we call it today, p. 24) and his calculation of areas with a curved boundary (p. 31), then again in great detail in Bolzano (pp. 104 and 105).

These authors did not create a name for their new object. (Others did, e.g. d'Alembert.) Cauchy too, now gives it a name:

> When the values successively attributed to the same variable approach a fixed value indefinitely, in such a manner as to end up differing from it by as little as we will wish, this final value is called the *limit* of all the others. It is written
>
> $$\lim x = X.$$

Undoubtedly, the concept "limit" is difficult. It is as difficult as the concepts "convergence" and "continuity"! Cauchy does not worry about these difficulties. (Pedagogical abilities do not always go hand in hand with subject knowledge.) His contemporary and textbook author Johann Tobias Mayer (1752–1830) tackles this problem in his book *Complete Teaching of Higher Analysis*, published in 1818, in quite another way. Mayer does not use the concept "limit" at all (which does not prevent him from writing at some point about a "ratio of limits"), but he treats the subject on no less than 20 printed pages!

This topic occupies Mayer's mind to such a degree that he even introduces *a special sign* for it. Interestingly, it is not the operator "lim" but the binary relation "≡":

> If one were to introduce a special sign in analysis to indicate an infinite approximation of some quantity to another, e.g. the sign ≡, nobody would take offence that, if the equation read
>
> $$T = \frac{1}{\log x} + \frac{1}{x^2} + \frac{1}{x^3} + \frac{1}{2^x}$$
>
> where the quantity x increases without end, i.e. becomes infinite, we would get
>
> $$T \equiv \frac{1}{\log x}.$$

As we remember, Johann Bernoulli did not come up with the idea of *a second sign for another type of equality:* p. 61. Times have changed. (As a mathematician, Mayer cannot hold a candle to Johann Bernoulli!)

Cauchy needs less than four lines in his preface for the same fact. This is followed by another four to five lines with two examples: irrational numbers as limits of fractions and the circle as the limit of inscribed polygons. His students will have found this hard to swallow, but this is quite another matter.

The Unspoken Luxury Version of the Concept of Limit

Cauchy's first example for his concept of limit, which was just mentioned above, is that of *the irrational numbers as limits of fractions.* However, there is a snag in this idea! *Strictly speaking,* Cauchy is not allowed to *introduce* an irrational number as the limit of an approaching sequence of fractions: it does not meet his notion of limit at all!

Just picture the problem: we do know what fractions are and we want to know what irrational numbers are. For us, at this stage, *the irrational numbers do not yet exist at all.*—Let us reread Cauchy's text: "When the values successively attributed to the same variable approach a fixed value indefinitely ..." It clearly mentions a "fixed value" which is to be "approached" by a sequence of values. But if the irrational number does not yet *exist* (quite to the contrary, it is yet to be constructed), then Cauchy's concept of limit is of no help—for there *does not yet exist* this "fixed value" which is to be approached by this sequence of fractions. This "fixed value" has, first of all, to be *created* by this construction.

However, Cauchy does not seem to be aware of this problem. *Directly afterwards,* just after having given his definition, he puts forward the irrational numbers as examples of "limits".

And this is no slip, for later on when he defines the operations for the calculation of the numbers *in detail,* he proceeds in the same way. Cauchy takes yet again the irrational number B just to be $B = \lim b$ where the (variable) quantity b is only allowed to have fractional values.

Today we have a problem with this strategy. We strongly discriminate between whether the "limit" spoken of already *exists* and whether it does *not yet exist* and is only to be *constructed* or *defined* with the help of this sequence. For us, these are *two different* notions of "limit". In the first case already existing facts are described: a given number is *called* a "limit", whereas in the other case a new mathematical entity is created through *being* a "limit". This second case could be named the "luxury version" of the notion of limit for it offers more than the other one, it *produces* a new mathematical entity (e.g. an irrational number).

In the true sense of the luxury version, an "irrational number" *is* nothing else than just this sequence of fractions, for which it is the "limit" (which is approaching it *as nearly as we wish*).

They *exist only* in this way. For example, π is *nothing else than* the sequence:

$$3; \quad 3.1; \quad 3.14; \quad 3.141; \quad 3.1415; \quad \ldots$$

Cauchy's contemporary Johann Traugott Müller was well aware of this (see pp. 200f).

What Is the Difference?

Cauchy writes as if there were no difference between the two (for us today) different versions. Was he conscious of this cheating?

I do not know. I did not find any formulation by him showing any awareness of this problem.

Let us remember Bolzano's view of this problem (see p. 104). According to him, a number is already *well-defined* if it could be calculated "as precisely as one wishes". As his contemporary, Cauchy could have thought in *just the same* way.

If this were the case, then for Cauchy the difference of these two versions of the notion of the limit was not discernible: it is the difference between those two "steps" which we made in presenting the modern notion of convergence (following Michael Spivak: p. 103)! We can state:

> Neither the brilliant thinker Bolzano nor the outstanding mathematician Cauchy in the first third of the nineteenth century made this distinction. However, Johann Traugott Müller did in 1838.

Consequently, this difference did not exist in their analysis. (But it *exists* today in our analysis.)

Some may judge this to be wrong. They might say that this difference between these two versions of the notion of limit (as well as the notion of convergence) is *eternal.*

But the proponents of this view are obliged to spell out what is meant by the "existence of a notion"—when even the most reknown mathematicians did not realize it. The hidden implication of this view is that mathematical notions (and truths) exist independently of their discovery and even if they are *not used* at all in *what is well-known* of mathematics.

"Function" and "value of a function" in Cauchy

Let us return to the foundational concepts of analysis, to one of the central notions. What is Cauchy's understanding of "function"?

The Concept of Function in Cauchy

Cauchy expresses very clearly and at length what he understands by a "function", namely a "changing" quantity:

> When variable quantities are so related among themselves that, the value of one of them being given, we are able to deduce the values of all the others, we usually consider these various quantities expressed by means of one among them, which then takes the name of the *independent variable,* and the other quantities, expressed by means of the independent variable, are what we call *functions* of this variable.

That is to say, given the four quantities x, x^3, $5e^{x+2}$, and $\sin x$, the last three are called "functions" of the first, which is the "independent variable" x.

Cauchy is very scrupulous and defines the same again—in case there are *more than one* "independent variable" (as in $5x^2 + 2y^3$). We can pass over this here.

The New in Cauchy's Concept of Function and a New Style of Notation

His detailed talk of "values" is easily discernible: the "independent variable" takes on "values" and from those the "values" of the variable quantity are "to be deduced". The whole is called a "function".

Not one mathematician before defined "function" in this way. This is definitely *new:* such detailed talk of the "values" of the variables. (We remember Bolzano's definition, which was given half a generation later: p. 111.)

Besides, a new entity deserves a new notation! Since Leibniz (and in the tradition of an idea from Descartes), the "variables" in mathematics are usually written as *small* and *italicized* letters, mostly "x", "y", "z". Now Cauchy introduces new and *important* entities. Consequently, it is utterly worthwhile that he introduces a new notation for these entities ("values"). (He clearly asks for this in his *Introduction!*) *Thus Cauchy consistently denotes the "values" of a "variable" in one of the following two ways: either he adds a* lower index *(e.g. "x_0" and "x_1" denote "values" of the "variable" x) or he uses the same letter written* in capitals and upright: "X", *too, denotes a "value" of x for Cauchy.* To sum up,

> In each case, Cauchy demands for an "independent" variable the precise specification *which* "values" it is allowed to take.

In general, Cauchy writes this: "Suppose x between x_0 and X." This means that the "variable" x takes *all* the "values" between the "values" x_0 and X. Today we write this as "$x_0 \leqq x \leqq X$" or in the language of sets "$x \in [x_0, X]$".

We will still have to think about the formulation "we are able to deduce" in his definition of function, but we postpone this to the section after the next. Let's ask first: is there anything else new with Cauchy's concept of function?

Cauchy's Concept of Function Is as Conservative as Possible for a Revolutionary

Apart from introducing the "values" of the variables (and consequently insisting on having a *restricted* domain for the "independent" variable), Cauchy's concept of function does not differ from that of Euler!

This we can *guess* from a comparison of the wordings of both definitions.

This can be *seen* in Cauchy's analysis: all his functions are given as *formulae—just as with Euler!*

And it can be *understood* by comparing Cauchy's definition with that given by Bolzano (some ten years later, p. 111). Bolzano operates *as generally as possible—* Cauchy *stays as closely as possible* to Euler's analysis.

In fact, Cauchy confines himself to transforming the "Calculus of Expressions" ("Algebraic Analysis") of his forerunners as closely as possible into his "Calculus of Values" (or "Analysis of Values"). He did not try to go beyond the intellectual horizon marked by the "Calculus of Expressions".

Considering the undisputed revolutionary nature of Cauchy's departure from the "Calculus of Expressions", one may judge this adherence to tradition to be appropriate or maybe even a clever diplomatic move. However, compared with Bolzano's way of thinking about analysis *and especially about the concept of a function*, Cauchy's thought is ultra-conservative—just as his personal convictions and his ideology.

Cauchy's Concept of the Value of a Function

Let us now return to the formulation in regard to the concept of a function which we have mentioned earlier: the "we are able to deduce". What could be meant by: we "deduce" the value of a function?

The problem is clear-cut. A "function" according to Cauchy (as well as to Euler) is e.g. $\frac{1}{x}$. Which values are there to be *deduced*? Clearly, all those which result from the instruction "Divide 1 by …". This is simple—only $x = 0$ is excluded, for division by 0 is not possible. So what?

In this case, Cauchy is *forced* to *define* what is to be done—what the "value" of the function *in such a case of exclusion* ought to be.

And Cauchy does, of course! He is completely explicit and declares:

> If a particular case arises in which the given function cannot immediately give the value of the function under consideration, we seek the limit or limits towards which this function converges as the variables approach indefinitely the particular values assigned to them. If there exist one or more limits of this kind, they are regarded as the values of the function under the given hypotheses; however many they may be. We call *singular values* of the proposed function those values determined as we have just described.

Cauchy's Concept "value of a function": A First Example

We consider the function $\frac{1}{x}$. If $x = 0$, we do *not immediately* obtain a value. *Therefore* we will look for *all* the "limits" which we can deduce from $\lim x = 0$:
$$\lim_{x \to 0} \frac{1}{x} = ?$$

This is not difficult: the smaller the value of x becomes, the larger the value of $\frac{1}{x}$ will be. And the sign of the value of $\frac{1}{x}$ is the same as the sign of x; both signs are possible. Cauchy writes an arbitrary large number (in other words, an infinite number) in the same way as Euler: "∞".

> Besides the numbers, ∞ is a "value", too.

The result is $\lim\limits_{x\to 0} \frac{1}{x} = \pm\infty$. —Alright?

A Surprise: Cauchy's "limit" Is Not Unambiguous!

This example (it is given by Cauchy) shows that Cauchy's "limit" is not in the least uniquely defined! It is possible to get *different* "limits" by *different selections* of values of the independent variable(s). In other words,

> Cauchy's concept of limit *generalizes* the concept of convergence.

Only in the case when the "limit" is uniquely defined, do the notions of limit and convergence coincide (see p. 110). And because "convergence" for the discrete is analogous to "continuity" for the continuum, we have:

> *The unique "limit" is a new formulation of "continuity".*

The notion of "limit" highlights another aspect of "convergence". (It denotes the *aim* rather than the *way* to that aim.)

I have to confess: there has been a little bit of cheating in my argument! It presupposed that ∞ was a "limit". This is *commonplace,* also in textbooks of analysis, for it is very convenient.

However, the assumption is *false!* The "value" ∞ has different properties from the "numbers"! *At no time* does a *finite* "value", an ordinary number, approach the "value" ∞ as close as one wishes—which is indeed the very definition of the concept "limit"! Quite to the contrary, the distance of some number a from ∞ is *always infinitely large:* $\infty - a = \infty$.

Nevertheless, it is common practice to take $\lim\limits_{x\to 0} \frac{1}{x} = \infty$, for some positive x, to be *really true.* In other words, when one considers the strict meaning of the concept of "limit", this equation is not *really true.* Instead, it is a—very convenient—*agreement.* Let us also stick to it.

A Second Example Relevant to Cauchy's Concept "value of a function"

For a better understanding of Cauchy's concept "value of a function", let us consider the sine function as a second example. We may visualize it by the following picture, a "curve" in the coordinate system:

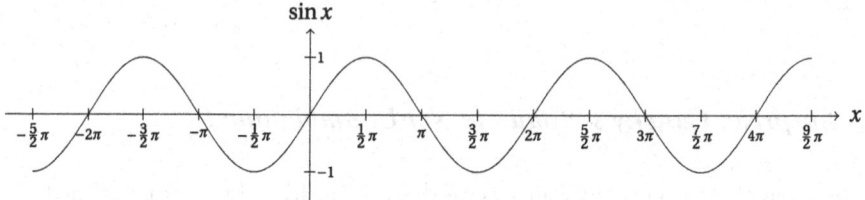

The sine function

The "value of the function" $\sin x$ takes on every number between -1 and 1 for all the "values" of x taken from any interval of length 2π. The "value of the function" $Y = \sin X$ can be determined by a computer program (today's style) or looked up in a table (old-fashioned)—or computed with the help of an infinite series (Leibniz-style).

According to our agreement, ∞ is a "value" too. Therefore, we may ask now: what is the "value of the function" $\sin x$ for the "value" $x = \infty$?

$$\boxed{\text{What is } \sin \infty?}$$

The answer is not too hard, or is it? It is *each* value from -1 to $+1$.

A short proof: let Y be any value in the interval $[-1, 1]$.

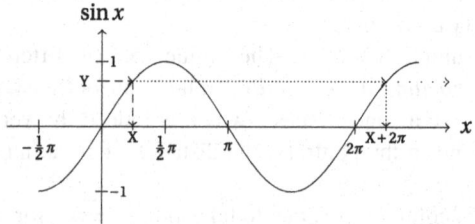

The value of the sine function for $X = \infty$

We are sure that there exists a value X between $-\frac{1}{2}\pi$ and $+\frac{1}{2}\pi$ with $\sin X = Y$.

(For this is a property of the sine function.) If $Y = 0$, we also have $X = 0$, for we have $\sin 0 = 0$. If $Y = +1$, we have $X = \frac{1}{2}\pi$, for we have $\sin \frac{1}{2}\pi = +1$, etc. And we also have $\sin(X + 2\pi) = Y$, for 2π is the period of the sine function. Consequently, we have $\sin(X + 4\pi) = Y$, too, and so forth.

$$\sin(X + 2k \cdot \pi) = \sin X = Y , \qquad \text{for all natural numbers } k.$$

Now we let k become infinitely large: $k \to \infty$, and consequently $2k \to \infty$ as well as $2k \cdot \pi \to \infty$. This amounts to

$$\sin\left(\lim_{k \to \infty}(X + 2k \cdot \pi)\right) = \sin(X + \infty) = \sin \infty = Y.$$

However, Y was *any* arbitrary "value" between -1 and $+1$. *Hence,* $\sin \infty$ may be *any* of these values from -1 to $+1$ too. If we write *all values* from -1 to $+1$ as the interval "$[-1, +1]$", we get our result as a simple formula:

$$\boxed{\sin \infty = [-1, +1]}$$

And this was to be proved! (Cauchy invents another symbol for this, see p. 147, but it means the same.)

Some Very Surprising Consequences from Cauchy's Concept of "value of a function"

The Methodical Significance of Cauchy's Definition of This Concept

A crystal-clear definition of the concept "value of a function" is an absolute must for Cauchy. After all, he did announce: "in determining these conditions and *these values*, I will make all uncertainty disappear". Since Euler, "function" has been the central concept of analysis, the *most important entity* whatsoever. *Therefore,* Cauchy has to declare *unmistakably* which "values" the "function" ought to take without ifs and buts.

Cauchy has met this requirement entirely. We have already examined his definition of the concept "value of a function" (pp. 124f).

The Historical Significance of Cauchy's Definition of This Concept

We emphasize:

> Cauchy is the first mathematician to define the concept of "value of a function".

From today's perspective, that seems to be surprising. However, if one is a little bit familiar with the kind of analysis *before* Cauchy, if one knows that the "Calculus of Expressions" set the academic standards before this time, the surprise will disappear. Those who know the "Calculus of Expressions" only a little, will be fully aware that the concept of value was of negligible relevance, playing only a minor role.

Cauchy is the very first mathematician who *explicitly* introduces the concept of "value" into the canon of fundamental concepts of analysis. This is a legacy of his long practice as an engineer during which Cauchy noticed: without numbers (as "values"), the algebraic expressions are of no use! This experience forced him to include the notion of value, which had been neglected until then, and place it at the centre of his theory.

The Political Significance of Cauchy's Definition of This Concept

Cauchy's manner of determining the concept of "value of a function" is of the greatest interest.

What does he say? This: the "value of a function" is *everything that is demanded by the existing conditions of the situation.* Cauchy says: if the situation is not *as clear as daylight by itself* (which means if the value cannot easily be calculated directly), we must take *all imaginable and* FITTING *values.* This is contrary to Bolzano's arbitrary requirement for all the values to be *somehow conceivable*—instead of only choosing those which *fit* to the *given* (existing) conditions. These are the "limits", *all of them.* Cauchy says: if we are not able to calculate the required "value" directly, then we *must exhaust all possibilities of the* EXISTING *conditions completely*—neither more nor less. This is still the generality of Euler's approach.

Do not question the existing limits of thought—this could threaten the order of the world! A greater contrast to Bolzano can hardly be conceived.

The Technical Significance of Cauchy's Definition of This Concept

Cauchy's definition of the concept of "value of a function" is also *in a purely technical sense* of greatest interest, for it is *of mathematical substance.* In other words, it allows for a very simple but powerful proposition, namely the following theorem:

Fundamental Theorem of Functions. *If the function* $f(x)$ *with the independent variable* x *has a unique finite value for the finite value* $x = X$, *then it is* continuous *at this value—and conversely.*

We learnt from Bolzano what continuity in the "Calculus of Values" meant (p. 108). Cauchy has precisely the same notion.

On p. 125, we already saw that the *unique* "limit" is the same as "continuity", and this is expressed exactly by the content of this theorem.—

A function is "continuous" for each value that is directly deduced from a calculatory expression. Today, the technical proof of this elementary fact is shown in every textbook of analysis. Its main idea is based on the fact that all values can be determined from a calculatory expression—as the operations are continuous.

Let me add two remarks.

1. The name of this theorem is new. It seems to be suitable.
2. Until today, nobody else has formulated this theorem, not even Cauchy.

What is the reason for that? Why did nobody formulate (and prove) this theorem which is true in Cauchy's analysis until today? Let me name three aspects.

(a) Within Cauchy's analysis, the theorem is *trivial*, a *mere truism*. One only has to recall the *strict analogy* of the concepts "convergence"—resp. *unique* "limit"— and "continuity". However, something which is trivial does not constitute a "theorem" for Cauchy!—Nevertheless, there is a formulation by Cauchy in a letter which is *exactly* the proposition of the theorem (letter to Coriolis from 13. February 1837).

(b) As all mathematicians might have noticed already,

> In today's analysis this theorem is wrong!

The reason is that in today's analysis, we use a concept of "value of the function" which coincides with that of Bolzano. As Bolzano's concept of "value of the function" allows for a completely arbitrary determination of the value of a function, this value *cannot* have any regular properties or nature. Therefore the "Fundamental Theorem" *cannot* hold within the conceptual framework of modern day analysis. It is just the *conservative* character of Cauchy's analysis which makes this theorem valid—more precisely, it is the stipulation of Cauchy that the "values of the function" have to be *deduced* and that they are not allowed to be chosen *arbitrarily*. It was only due to the revolutionary Bolzano who mandated this, and we follow this conceptual framework until today, with all the consequences.

(c) Taking both points into account we can conclude: not necessarily Cauchy, but all those mathematicians, *who deal with Cauchy's analysis today,* have every reason for stating this theorem. It *shows explicitly* that and in which way analysis *has changed* from Cauchy until today. The mathematicians who deal with Cauchy's analysis today are the historians of mathematics. It is *they* who have to supply the information why they still refrain from pointing out this fine theoretical change. Since the 1990s (when I published my interpretation of this theorem), they have not risen to the challenge. More precisely, they ignore these facts. (This in turn raises the question of the current state of the history of mathematics as a science: what is its character today? Are the historians of mathematics afraid of mathematics?—We shall return to this.)

Excursus: Preview of a Failed Revolution of Analysis in the Years of 1958 and 1961

In the years 1958/61, analysis had the chance to revolutionize itself, to undergo a conceptual upheaval comparable with the one caused by Cauchy. It was proposed to found the new analysis on a *different concept of number.*

These ideas were published by two German authors and, quite independently, by an American author. The latter was born in Germany to Jewish parents who went into American exile in order to escape German fascism. He later became a mathematics lecturer in the USA.

Cauchy's example shows clearly that the detailed formulation of any analysis also depends on number and, consequently, on the accepted concept of number. The consequence is this:

> The introduction of *another* concept of number—i.e. of numbers with *other properties* than those used by Cauchy and ever since Cauchy—would change the configuration of analysis. The form of the "Analysis of Values" *would have to change* accordingly.

This was exactly the issue when in 1958 the article *An Enlargement of Calculus* by Curt Schmieden (1905–91) and Detlef Laugwitz (1932–2000) and in 1961 the paper *Non-Standard Analysis* by Abraham Robinson (1918–74) were published. Whereas Schmieden/Laugwitz spoke of a "proper enlargement" of calculus, Robinson declared a "proper enlargement of classical calculus" and at once came up with the name for his new theory: Nonstandard-analysis.

History Does Not Recur, Not Even in Mathematics

But 1958 was not 1821. One and a half centuries after Cauchy's textbook, it was *plainly impossible* to reorganize mathematics as completely as Cauchy had done (at least in large parts, with the exception of one aspect which will be mentioned on pp. 135f). Analysis was established far too deeply, in thousands of heads and hundreds of textbooks.

In 1821, the formation of analysis was the duty of a few prominent authorities. In those days, active mathematicians who dealt with analysis knew of each other. Nearly all of them lived in Paris or went there temporarily to do their years of apprenticeship. But by 1958, analysis was a worldwide established subject, at all scientific colleges and universities. Courses and syllabuses existed as well as textbooks and teachers—who were all *trained to think in the same analytical way.* Should all of them *change their minds?* Why? Who could direct or initiate this? Who should execute or control this? Such an expenditure, and for what purpose?—The traditional analysis was not *wrong* or *impracticable.* At worst, it was constructed in a somewhat complicated way (see Chap. 14). This is one of the main reasons used by the proponents of the nonstandard-analysis to persuade others to accept their theory. But *only to change the way of thinking,* maybe to simplify it, and this without producing new results? Of course, this is not enough to overthrow a worldwide church—moreover in a field where the mindset is canonized in all detail in an unparalleled manner. It would have required at least a new truth in order to give such a revolution at least a tiny chance of success. However, nonstandard-analysis did not offer any new truth—but only the old in a new guise. The hope of some pioneering rebels to persuade by pointing to *the existence of some new mathematical subjects* (e.g. to delta-functions, even rational ones!—see p. 240) remained unfulfilled. Delta-functions are a fascination only for specialists.

Under these circumstances a new revolution of analysis in the last third of the twentieth century did not occur. The tanker *Analysis* had grown far too sluggish and big and could not be *reversed* by individuals or by small groups.

> As soon as a system of thought is broadly established as well as institutionalized in a society—and in this way becomes *universally valid*—, it is impossible for a group of persons to take over. Majority is power when there is no absolute rule.

To put it ideologically, this is not a question of mathematical *truth* whatsoever. Nonstandard-analysis is neither more nor less "true" than standard-analysis (as the commonly used form of analysis is called now, for the sake of precision). Just as Cauchy's "Calculus of Values" is neither more nor less "true" than the "Calculus of Expressions" from Euler and Lagrange. The "Calculus of Values" is *more useful* in respect to the actual needs of the time (technical usage) than the former construction. However, it is not *"more true"* in any possible sense. The proofs of its theorems are by no means *more precise* than the proofs of the earlier theorems.

Unfortunately, historians of mathematics very often state the opposite of this, today more than ever before. (Exceptional authors like Henk Bos or Kirsti Andersen only confirm the rule.)

Of course, this is nonsense! Why should we, today, be qualified to think more precisely than our ancestors? There is no obvious reason for this. Majority is not law (or truth). Those who judge in this presumptuous manner are either too lazy to familiarize or not capable of familiarizing themselves with *another* way of thinking—as we are attempting in this book.

A Rebellion of Nonstandard-Analysis?

Many articles dealing with Cauchy's analysis suddenly appeared in the 1970s and 1980s. The reasons for this can only be guessed. One hypothesis is that this hype was a last attempt by the proponents of nonstandard-analysis to justify their theory by trying to place it within a historical tradition. And this in place of offering some new truth: history instead of mathematics?

The idea was: if it could be Nonstandard-analysis is the "true", the "right" understanding of analysis, then it would show its superiority—at least from a moral point of view.

This line of argumentation drew on history. The master plan was to *prove that some first-rate mathematicians were* ESSENTIALLY *doing nonstandard-analysis!*— an idea, no doubt.

The first step was very easy. *All* proponents of analysis living before the twentieth century liked to use the attribute "infinitely small". However, this was exactly the novelty of Nonstandard-analysis, its characteristic distinction from standard-analysis:

In Nonstandard-analysis there do exist "infinitely small" (and similarly also "infinitely large") *numbers*—which do not exist in standard-analysis.

(We will return to this, see pp. 226f, especially p. 227.)

That is why at first sight it was easy for the nonstandard-analysts to live off the reputation of their famous forerunners. In respect to Leibniz, the fine details of his concepts had still not been published, and his phraseology "infinitely small" could therefore be interpreted very freely: namely in a modern nonstandard-analytical way. With Euler and Johann Bernoulli this could be done effortlessly, for they *really* talked about such "infinite" *numbers*—we have read this! Consequently, they could easily be *co-opted* as being *essentially* proponents of Nonstandard-analysis. It did not matter here that neither Euler nor Johann Bernoulli made any attempt to explain how such a concept of number could be *explicated:* as *really great* mathematicians *must think in the right way*—there should be no problem!

Obviously, this practice deviates completely from doing history of science, as it is undertaken here. (By the way, even today there exist fundamentalists of history of mathematics who are unable to understand this way of thinking. A rational discussion with them is completely impossible. —Do you imagine such a controversy *in mathematics?*)

It was of great help when philosophers started to join the debate. Thus, the philosopher of science Imre Lakatos (1922–74) subscribed to the opinion that Cauchy's style of thought was a nonstandard-analytical one. Lakatos' in some way risqué arguments could be turned into merely stringent ones (which helped me to some renown after having written a book on it). As a result, the youngest of the triumvirate which started Nonstandard-analysis in the years 1958/61, Detlef Laugwitz, took the plunge in construing Cauchy as a *true* nonstandard-analyst, who actually thought within these concepts. Laugwitz succeeded in giving many beautiful mathematical arguments, but he failed in his *intent,* for:

The interpretation of Cauchy's "Calculus of Values" as an (early) state of Nonstandard-analysis can be proven to be false.

The reason of this will be explained shortly.

After Leibniz, Euler and Johann Bernoulli—as just stated—, there still remained Cauchy. Needless to say, Cauchy also uses the attribute "infinitely small". Yet, there is something else!

A Digression Within the Excursus: Looking Back at a Criticism of Cauchy

Five years after Cauchy's textbook of analysis appeared, a claim of the young mathematician Niels Henrik Abel (1802–29) was printed which says that Cauchy's book contains a false theorem. (This will be spelled out in some detail on pp. 140f.) However, there was no direct reaction from Cauchy to this reproach.

Consequently, the analysis of the year 1826 was faced with a factual problem: is this theorem true or false? Cauchy had stated and proved it—Abel had pronounced it wrong. Two nowadays still very much appreciated mathematicians contradicted each other in their judgement.

24 years later, a treatise from Philipp Ludwig Seidel (1821–96) was published which also criticized that theorem of Cauchy. It also stated and proved an alternative theorem. Again, there was no reaction from Cauchy.

However, three years later (in 1853), Cauchy published a treatise which contained—in some hidden way, which we will examine later—his self-defence in this matter. Because Cauchy did not *explicitly* label this statement to be a "self-defence", it remained unnoticed. Only much later, when mathematicians finally realized the significance of Cauchy's treatise of 1853, they started to discuss it (p. 140).

Back to the Upheaval of Nonstandard-Analysis

With the birth of Nonstandard-analysis 1958/61, a quarrel had started which finally reached a climax in the 1970s and 1980s. Today this controversy is completely over and this to such an extent that it could not offer any contribution to a new idea concerning Cauchy's theorem. The controversy is:

Which of the two alternatives is true?

1. Cauchy stated and proved in his textbook a (very elementary) false theorem.
2. Cauchy's theorem is true—because by "convergence" Cauchy meant not the same as we do today, but instead he meant what we call nowadays "uniform convergence".

The first alternative is not plausible at all. Inevitably, every mathematician makes mistakes. But a mistake such as this? A mistake regarding a theorem studied in the first course of analysis—which did not lead to a proper defence from its author, and this even after the publication of a criticism? Who should believe in this unlikely narrative?

Subsequently, the nonstandard-analyst s felt newly empowered: they simply changed the interpretation of the decisive concept of this theorem from Cauchy's (for him it is "convergence") to mean now "uniform convergence"—and that the theorem *in this new meaning* is correct is disputed.

However, the nonstandard-analysts clearly remained a clear minority. Therefore, the alternative (1) remained the dominant opinion and Cauchy became the most popular half-wit in the textbooks of analysis.

Walking on Very Thin Ice

The question for the nonstandard-analysts was: how could they justify changing the interpretation of Cauchy's concept? Their idea showed great promise. They argued: *Cauchy did not understand by "number" what everybody else (for at least one hundred years) thought of the concept but, quite contrarily, what we, the nonstandard-analyst s mean by it* TODAY!

From the mathematical point of view, they had succeeded. The reason being: the range of numbers within Nonstandard-analysis is considered to be definitely greater than the range of real numbers, say: the decimals (however, see chapter 14). And in the theorem under consideration Cauchy makes a certain supposition: he demands that a "series" is "converging" *for all values* (and he meant for all *real* numbers). However, if by the "for all numbers" he meant *more* than we thought until now (namely besides the "real" numbers also the "hyper-real" numbers), then this surplus in the assumption allows for a surplus in the conclusion, and that is, indeed, the claim of the theorem.

This mathematical argument is indubitably correct. Robinson had published this reasoning already in 1963, and Lakatos, this ingenious propagandist of his own ideas, exaggerated its importance, especially in philosophy of science. (To be honest, once I joined in this enterprise.)

Yet, the decoding and interpretation of former mathematical concepts is neither the duty nor a self-evident competence of a pure mathematician! Instead, it is a main task for the historians of mathematics.

However, a detailed historical investigation of the concepts used by Cauchy within his analysis was not forthcoming. Nobody undertook an inquiry of Cauchy's concept of number to check whether he *really* considered such exotic entities like "infinitely small" *numbers*. And, of course, Cauchy did not do that!

In any case, Nonstandard-analysis needs to perform some acts of conceptual acrobatics in order to construct "hyper-real" numbers in a mathematically acceptable way. Here Robinson's construction stands out, but without studying (at least) one semester of modern logic, one is not able to follow his construction. It is, however, unlikely that Cauchy should have anticipated such a stilted concept in 1821 in any possible sense. For this, we need a further supporting argument! But nobody has supplied one: instead there were plenty of rhetorical fireworks!

In short, this attempt by some nonstandard-analysts to justify their theory by recourse to the development of analysis remained without historical substantiation. (This is true, although at the time of my dissertation, around 1980/81, and some years later, I was convinced of the opposite—and so were the referees of my thesis, who were mathematicians.)

Not earlier than 1990, I undertook a thorough examination of those concepts which Cauchy *really* used. My result can be summarized as follows: if you have to choose between two possibilities, just take the third! Or, to put it somewhat more technically, the translation of Cauchy's concept of "convergence" in modern analysis is neither "convergence" nor "uniform convergence"—but a third notion. This will be explained in the next section.

Cauchy's Concept of Convergence: A Big Misunderstanding

Strictly speaking, the discussion of Cauchy's concept of convergence belongs to the section which is called "some very surprising consequences from Cauchy's concept of 'value of a function'", but because of its huge importance it will be considered separately.

A Mystery of History: Cauchy's Concept of Convergence

There is no other concept of analysis which caused such long, comprehensive, and vehement controversies and which also went beyond the field of history of mathematics, than Cauchy's concept of "convergence". Why is this?

The decisive point is: although Cauchy's analysis is shaped very revolutionarily, it remains firmly tied to established thinking—which is thoroughly alien to us today.

Let us return to Cauchy's concept of function (p. 123). Which basic notion does he use? It is "quantity". We should bear in mind:

> Cauchy's first and unique foundational concept of analysis is "quantity". (From that he also deduced the concept of "number".)

As a consequence, Cauchy determined "convergence" for those objects, exclusively for quantities.

> Originally Cauchy's concept of "convergence" applies to "quantities".

(To overlook this has been the decisive mistake in the mathematical–historical debate following the creation of Nonstandard-analysis and its efforts to co-opt Cauchy since the early 1960s. However, *today* no mathematician is able to give an account of this concept: "quantity". Cauchy's definition of "quantity" can be read on p. 118.)

Here is Cauchy's definition of "convergence":

We call a *series* an indefinite sequence of quantities,

$$u_0, \quad u_1, \quad u_2, \quad u_3, \quad \ldots,$$

which follow from one to another according to a determined law. These quantities themselves are the various *terms* of the series under consideration. Let

$$s_n = u_0 + u_1 + u_2 + \ldots + u_{n-1}$$

be the sum of the first n terms, where n denotes any integer number. If, for ever increasing values of n, the sum s_n approaches a certain limit s, the series is said to be *convergent,* and the limit in question is called the *sum* of the series.

The u_k are "quantities", e.g. "functions" $f_k(x)$ (Cauchy still uses the comma in its traditional meaning of the plus sign):

$$f_0(x) + f_1(x) + f_2(x) + f_3(x) + \ldots$$

We realize that, in the case of series, Cauchy deviates from his usual way of signification; in this case, the index does *not* indicate a "value" but instead belongs for practical reasons to the name of the "quantity". (Here the name "f"—in Cauchy: "u"—, supplied with an index, does not denote a variable! It is merely a "variable" *name*.) In regard to Euler and Lagrange we have already seen how difficult it is to deal with series without the index notation: pp. 75 and 95.

Starting from a "sequence", Cauchy now constructs *finite* sums $s_n(x)$:

$$s_n(x) = f_0(x) + f_1(x) + f_2(x) + \ldots + f_{n-1}(x).$$

If the $s_n(x)$ approach a "limit", this is called "convergence". Written in Cauchy's way, we have

$$\lim s_n(x) = s(x).$$

Before we proceed with the technicalities, we need to discuss an emerging problem: Cauchy speaks of a "limit". However, a "limit" is a value. Yet this $s(x)$ is a *quantity (as can be seen by the "x"), no value.* Notwithstanding, Cauchy *speaks* of the *"limit"!* The only possible explanation is that Cauchy means "the *value* of $\lim s_n(x)$ for a $x = X$"!

Consequently, Cauchy's definition of "convergence" is this:

A series of functions

$$f_0(x) + f_1(x) + f_2(x) + f_3(x) + \ldots$$

is called "convergent" for the value $x = X$, if for increasing n the sum

$$s_n(x) = f_0(x) + f_1(x) + f_2(x) + \ldots + f_{n-1}(x)$$

has for $x = X$ a unique "limit".

It is possible to reformulate this definition. Instead of "$\lim s_n(x) = s(x)$", we may write "$\lim(s_n(x) - s(x)) = 0$", couldn't we? Using Cauchy's notation for this difference, which is still in use today, "$r_n(x)$" ("reste" = "rest"), we get:

A series of functions

$$f_0(x) + f_1(x) + f_2(x) + f_3(x) + \dots$$

is called "convergent" for the value $x = X$, if for that value the special sum (the "rest")

$$r_n(x) = f_n(x) + f_{n+1}(x) + f_{n+2}(x) + \dots$$

we have

$$\lim r_n(x) = 0.$$

The Solution of the Mystery

Until 1990, there has been a consensus among all Cauchy specialists about this formulation. But since 1990, there has existed a new idea, arising from *two open questions:*

1. *What IS "$r_n(x)$" for Cauchy?*
   The only possible answer is a *"variable"*, depending on the TWO "independent" variables n AND x.

2. *What IS "$\lim r_n(x)$" for Cauchy?*
   Again the answer is simple: $\lim r_n(x)$ as a mathematical notion in Cauchy's analysis is a *"value of a function"* and, consequently, "the limit or the limits"— to Cauchy: *all* of them!—of the variable $r_n(x)$ (for the value $x = X$).

And that is the point! By "$\lim r_n(x)$ for the value $x = X$", Cauchy *not only* means

$$\lim_{n \to \infty} r_n(X),$$

but moreover, there are also *all* "limits" to be included:

$$\lim_{\substack{N \to \infty \\ x_k \to X}} r_N(x_k), \tag{$\|$}$$

for $r_n(x)$ is a function of the *two* variables n and x.

Cauchy *never* writes subscripts when he uses "lim". Only later did it become customary. The example above shows *why it was superfluous for Cauchy:* he always means *all* possible limits. Our example also shows that this notation—which today is a *must*—is generally somewhat laborious. Thus, as Cauchy usually means all limits, he can use "lim" unindexed.

The Mathematical Significance of This Solution

We remember Cauchy's theorem and that it has been considered false (except by nonstandard-analysts) until today. It reads literally (only the sign " X " is added two times):

Theorem I. *When the various terms of the series*

$$u_0, \quad u_1, \quad u_2, \quad \ldots, \quad u_n, \quad u_{n+1}, \quad \ldots$$

are functions of the same variable x, continuous with respect to this variable in the neighbourhood of a particular value X for which the series converges, the sum s of the series is also a continuous function of x in the neighbourhood of this particular value X.

In the special literature, sometimes the name "Cauchy's Sum Theorem" is used.

The decisive assumption of this theorem is that the series of functions "converges" and this also "in the neighbourhood" of the particular value X.

Today we can offer this assumption in the terms of all *three* rivalling interpretations:

1. The standard-analysts understand Cauchy using the concepts which are customary today. Which means they say " $\lim_{n\to\infty} r_n(X) = 0$ ".

2. The nonstandard-analysts understand Cauchy using their concepts and say " $\lim_{n\to\infty} r_n(X') = 0$ *for all values of* X' *in the neighbourhood of the value* X *in question.*"

3. According to the new interpretation, obtained from an investigation of Cauchy's concepts, we reach the conclusion that Cauchy means what is formulated in line ‖. (By "neighbourhood", Cauchy means "X + α" instead of other "values" X' ≠ X.)

Stated in the terminology of *modern* analysis, this amounts to the following:

1. In his theorem, Cauchy presupposes the "convergence" of the series and consequently the theorem is *wrong.*
2. In his theorem, Cauchy assumes the "uniform convergence" of the series and consequently the theorem is *true.*
3. In his theorem, Cauchy demands the "continuous convergence" of the series and consequently the theorem is *true.*

The *exact* definition of "continuous convergence" can be found in some textbooks of the theory of functions, published since 1921. For the understanding of the historical development, it is not necessary to understand this definition; it is enough to accept the following box.

In case (1), Cauchy looks bad. In case (2), the standard-analysts look bad. In case (3), after having investigated Cauchy's concepts, the historian of mathematics reaches the judgement: Cauchy did not make a mistake here, but he proves something other than the standard- as well as the nonstandard-analysts think.

In the terminology of modern analysis, we state the following ("$\Rightarrow$" means "leads logically to", and *not one* of the following inversions is true):

continuous convergence	$\Rightarrow$	uniform convergence	$\Rightarrow$	convergence

The *weakest* presupposition (on the right side) is not strong enough to prove the theorem. The presupposition *in the middle* is stronger and suffices—however, there is no clear evidence in Cauchy's text that he might have meant this. The *strongest* presupposition (on the left) is Cauchy's *true* condition—and it is all the more sufficient to prove his assertion.

Cauchy's Proof of His Theorem

We finish our argument by replicating Cauchy's proof of his theorem. The proof is simple.

The claim is that the sum $s(x)$ is continuous at a value $x = X$. Continuity means $s(x + \alpha) - s(x)$ decreases (for the value $x = X$) together with α and becomes arbitrarily small. This has to be proven:

$$\lim \left(s(x + \alpha) - s(x) \right) = 0 \qquad \text{for} \qquad \lim \alpha = 0 \quad \textit{and} \quad \text{for the value} \quad x = X.$$

This is a consequence which follows directly from the usual representation of s as $s = s_n + r_n$:

$$s(x + \alpha) - s(x) = s_n(x + \alpha) - s_n(x) + r_n(x + \alpha) - r_n(x). \qquad (**)$$

Since the s_n are finite sums of continuous functions, $s_n(x + \alpha) - s_n(x)$ is a finite sum of infinitely small quantities and, consequently, infinitely small. Only the two other summands remain $r_n(x + \alpha)$ and $r_n(x)$—not to forget, in each case for the value $x = X$.

However, we have $\lim r_n(x) = 0$ (for the value $x = X$): this is *exactly* the presupposed "convergence" of the series of functions for the value $x = X$.

Similarly, we know $\lim r_n(x + \alpha) = 0$ for the value $x = X$, *because of Cauchy's concept of "value of a function"* —here for the value X! For we know "$f(X)$" are *all* limits $\lim f(X')$ for $X' \to X$, and this can be written as $\lim f(X + \alpha)$ for $\alpha \to 0$, or just for our function $r_n(x)$: $\lim r_n(X + \alpha)$ for $\alpha \to 0$. This is required to be $= 0$ *because of the assumed "convergence"—in Cauchy's sense!*—FOR THE VALUE X. End of the proof.

The very last step has always been controversial:

1. The standard-analysts read "$x + \alpha$" as VALUES *different from* X and say Cauchy cheats by assuming "convergence" for values *different* from the value X—but this is not permitted. However, for Cauchy "$x + \alpha$" is *always* a "variable" and *never* a "value"!

2. The nonstandard-analysts argue no problem, α is an "infinitely small" *number*—and these are included in Cauchy's presupposition of "convergence" *even in the neighbourhood*. And this fits!—however, not to Cauchy's way of thinking.

It is impossible clearly to demonstrate the concept of "infinitely small" *number* within Cauchy's texts. But Cauchy offers a concept of "value of a function" (in detail, *all* the limits!) nobody had ever thought of. Today this concept is unknown, and before 1990 nobody read Cauchy's textbook *very carefully*.[1] And because of Cauchy's concept of "value of a function", "$\lim r_n(x+\alpha) = 0$ for the value $x = X$". This is guaranteed by the presupposition of "convergence (at the value $x = X$)". —Bingo!

Cauchy's Self-Defence

As stated earlier, Cauchy later published a treatise in which he answered his critics. He did so in a very polite manner and did not name them (nor their criticism) explicitly. Niels Henrik Abel had criticized Cauchy's theorem in all detail in 1826 (see p. 133). Abel declared the theorem to be wrong. However, he did not *prove* this, but merely *claimed* that a particular function is a counter-example (see p. 156).

In a treatise from 1853 Cauchy offers a calculation which *proves* Abel's claim to be false.

In place of Cauchy's somewhat demanding calculation,[2] we transcribe it and Abel's counter-example into a simpler version.

Abel says something like the following. The series

$$x^0 + (x^1 - x^0) + (x^2 - x^1) + (x^3 - x^2) + \ldots$$

converges for each value of x between 0 and 1 inclusively. Its single terms $x^{n+1} - x^n$ are continuous. However, the whole sum is not, and therefore it is a counter-example to Cauchy's Sum Theorem of p. 138!

Let us examine this argument. Abel is correct in saying that the terms $x^{n+1} - x^n$ of this series are continuous, but that the sum is not. Why the latter? If we let $0 \leqq x < 1$, this sum has the value 0, for we have

$$s_n(x) = x^0 + (x^1 - x^0) + (x^2 - x^1) + \ldots + (x^n - x^{n-1}) = x^n \, ,$$

and consequently (for $x = X$ and since $0 \leqq X < 1$)

---

[1] Not even the translators Robert E. Bradley and C. Edward Sandifer in 2009 did!

[2] See *Die Analysis im Wandel und im Widerstreit*, p. 352.

$$s(X) = \lim s_n(X) = \lim_{n \to \infty} X^n = 0.$$ (††)

However, for $x = 1$, the series is plainly one:

$$s(1) = 1 + (1 - 1) + (1 - 1) + (1 - 1) + \ldots = 1.$$

Subsequently, *this* function $s(x)$ is discontinuous for the value $x = 1$, because for $X < 1$ we have from ††:

$$\lim_{X \to 1} s(X) = \lim 0 = 0,$$

whereas we have $s(1) = 1 \neq 0$. In Abel's mindset, this disproves Cauchy's theorem.

Nevertheless, this does not follow from Cauchy's way of thinking! Quite contrarily: Cauchy is able to *prove* that this series of functions does *not* "converge" for the value $x = 1$—namely in the *special* sense in which Cauchy had established the concept of "convergence"! But if this series *does not meet the presuppositions of his theorem*, his theorem cannot state anything about it.

To conclude, we will show that our example of a series for the value $x = 1$ does not "converge" in Cauchy's understanding of that notion. This is quite easy. We only have to study $\lim r_n(x)$ for the value $x = 1$. We have

$$r_n(x) = (x^{n+1} - x^n) + (x^{n+2} - x^{n+1}) + (x^{n+3} - x^{n+2}) + \ldots = -x^n.$$

Is $\lim r_n(x) = -x^n$ for the value $x = 1$ also $= 0$?—No, for the value $x = 1$, we get $\lim r_n(1) = \lim -1^n = -1 \neq 0$. Quite simple!

Abel's counter-example (which is technically somewhat more complicated) cannot be refuted that easily. An additional consideration is needed. This additional consideration can also be explained by using our simple example (the original argument was given by Cauchy in 1853). It runs like this. Instead of taking $x = 1$, we investigate the $\lim r_n(x)$ for $x = 1 - \alpha$ and $\lim \alpha = 0$. All the values $1 - 1/n$ are < 1 and we have $\lim\limits_{n \to \infty} (1 - 1/n) = 1$. Therefore $\lim\limits_{n \to \infty} r_n(1 - 1/n)$ is also a "value of the function" $r_n(x)$ for $x = 1$. However, it is (compare with the formula on p. 77, for $k = 1$, we get $a = e$, and also, choose $x = -1$ *resp.* $i = -\frac{1}{\omega} = N$):

$$\lim_{N \to \infty} r_N(1 - 1/N) = \lim_{N \to \infty} -\left(1 - \tfrac{1}{N}\right)^N = -(e^{-1}) \neq 0.$$

A completely flawless calculation—which constitutes in Cauchy's analysis the *proof* that this chosen series, *in his understanding* does *not* "converge" for the value $x = 1$! (Cauchy only skipped the part "$\lim r_n(x) =$" in his calculation—this step he assumed to be clear to his learned readers.) Clearly, the meaning of Cauchy's calculation can *only* be understood if one relies on his concept of "convergence" *as well as*—and at least equally importantly!—his concept of "value of a function".

However, his critics have refused to do this: in the nineteenth, in the twentieth, and (until now) in the twenty-first century.

By the way, this example of a series (just like that which Cauchy had provided) is *not "uniformly convergent"*. That is why it (as well as Cauchy's series), sadly, is not suited to refute the claim of the nonstandard-analysts, that Cauchy had thought of his "convergence" in the same way as them, mathematically. That is a pity, because a *technical* argument would be considered by many as mathematically more conclusive than a merely *philosophical* one.

What Are the Reasons for the Prevailing Misunderstanding of Cauchy's Notion of Convergence?

Why do modern mathematicians not understand Cauchy's concept of "convergence"? The brief answer is: because they are no historians of mathematics. A more detailed answer consisting of two parts is given below:

1. Cauchy has defined this concept for "quantities"—whereas we (in analysis) define it *exclusively* for "numbers".
2. Cauchy's concept of "value of a function" is quite different from that which we use today (p. 124), a fact that has stayed unnoticed until now.

We have reached the result: if we take Cauchy at *his* word, the quarrel about his theorem which has now lasted for more than 96 years has finally been resolved. Strangely enough, today nobody is interested: each publication of this explication remained without professional resonance. Will this change now?

Today it is even possible to accuse Cauchy—by ignoring his conceptual limits— of *new "errors"* in his analysis and to publish this in a scientific journal viewed to be first rate, like *Historia Mathematica*. And even worse, the named journal refused to publish a criticism of such an "erroneous" article. (The editor then decided, relying on the—clearly contra-factual—argument, this journal would not publish "Letters to the Editor". However, I did write an article, not a letter, but it was not even reviewed.) From this, we may conclude that history of mathematics is not a separate subject, but only the servant of ideological warfare. It cannot be part of an independent scientific discourse.

Cauchy's Concept of Derivative—Again a Misunderstanding

The notion "derivative" was introduced by Lagrange. Lagrange defined it as a "function" $f'(x)$ which can be determined from the series expansion of the function $f(x)$ by changing the argument, i.e. from the series expansion of $f(x+a)$ (formula § on p. 96).

Instead of working with "derivatives", Euler operated with "differential quotients", which he took to be *true* quotients made from two "infinitely small" quantities. This is a complicated notion and was therefore avoided by Lagrange.

As we have read in the preface of his textbook on analysis, Cauchy made a point of "making all uncertainty" of the formulae "disappear" by the "determination of the values". Cauchy follows this principle also in case of the derivative. He introduces a formulation which is still, today, presented to the beginners—even though, again, in another meaning.

Cauchy refers to the fact that for a "continuous" function $f(x)$, an "infinitely small" increase α of the variable x (i.e. he takes $x + \alpha$ and demands $\lim \alpha = 0$) produces an "infinitely small" increase of the function. Cauchy likes to denote these "increases" by capital delta (Δ), i.e. $\alpha = \Delta x$ as well as $f(x + \Delta x) - f(x) = \Delta y$. We read Cauchy (by "the two terms", he always means numerator and denominator of the fractions):

By consequence, the two terms of the *ratio of differences*

$$\frac{\Delta y}{\Delta x} = \frac{f(x + \Delta x) - f(x)}{\Delta x}$$

will be infinitely small quantities. But, while these two terms indefinitely and simultaneously will approach the limit zero, the ratio itself may be able to converge to another limit, either positive or negative. This limit, when it exists, has a determined value for each particular value of x; but, it varies along with x.

Cauchy defines this "value" of $\lim \frac{\Delta y}{\Delta x}$ for $\lim \Delta x = 0$ to be the "value" of a new "function":

The form of the new function of the variable x, which will serve as the limit of the ratio $\frac{\Delta y}{\Delta x} = \frac{f(x+\Delta x)-f(x)}{\Delta x}$, will depend on the form of the proposed function $y = f(x)$. To indicate this dependence, we give to the new function the name of *derived function*, and we represent it, with the help of an accent mark, by the notation

$$y' \quad \text{or} \quad f'(x).$$

As already stated above, even today the "derivative" (as we call the "derived function" of Lagrange and Cauchy today) is introduced to beginners in this way. However, as Cauchy's thinking *basically* differs from that of ours today, we must add two things.

1. To say only the least, in Cauchy's way of thinking, this manner of defining a "function" is problematical because Cauchy establishes a "function" to be a "variable" which has certain properties (see p. 123). However, the "derived function" is *not* defined to be a certain variable, but it is constructed as a value-to-value relation. (At most one may, *in hindsight, call* that value-to-value relation, defined by Cauchy, a "variable".)

> *The manner of defining a "function",*
>
> $$X \longmapsto f'(X),$$
>
> *is very modern* (to be precise, this *is* the method we use today), *but it definitely does not fit to the concept of "function" given by Cauchy!*

In other words, by giving this notion of the "derived" function, Cauchy violates his own conceptual framework. Strictly speaking, in Cauchy's thinking, his "derived function" is no "function" at all!

2. What is Cauchy's *precise* definition of "derivation"? It is that he determines which is the (uniquely defined!) "value" $f'(X)$ for the "value" $x = X$—namely the value which is the "limit of $\frac{f(x+\Delta x)-f(x)}{\Delta x}$".

> The value $x = X$ produces the value $\lim \frac{f(x+\Delta x)-f(x)}{\Delta x}$ for $\lim \Delta x = 0$.

However, we must also take into account that Cauchy takes each value X (also) as a limit $\lim x_n = X$! In other words, Cauchy does mean the following:

> The value $x = X$ produces the (unique!) value $\lim \frac{f(x_n)-f(\bar{x}_n)}{x_n-\bar{x}_n}$ for $x_n \neq \bar{x}_n$, $\lim(x_n - \bar{x}_n) = 0$ (with $\lim x_n = X = \lim \bar{x}_n$).

Nonetheless, in today's language, this is the definition of *"continuous derivability"!* This entails that Cauchy's definition of derivability *demands* that the derived function $f'(x)$ is continuous. (And that even though his *formula* is the same as ours!—However, we already know that Cauchy's concept of "value of a function" differs from that of today—and this must have mathematical consequences. Mathematics involves not just formulae but also the *interpretation* of the formulae. In Cauchy's words: we must make the "uncertainty" of the formulae "disappear".)

After having realized that Cauchy's concept of convergence is equivalent to today's concept of continuous convergence, we shall find it evident that Cauchy uses also the modern concept of "continuous derivability" *instead* of our "derivability".

Let me add two remarks, a historical one and a mathematical one.

(a) The historical one: Laugwitz argued already in 1987 that Cauchy's "derivability" is our modern "continuous derivability". His argument was different and I was unable to accept it. Unfortunately, even in *Die Analysis im Wandel und im Widerstreit*, pp. 333–335, I had not understood Cauchy correctly.

(b) The mathematical one: Strangely, in today's lectures for beginners, it is not mentioned that the formula

$$\lim_{\substack{x_n \to x_0 \\ \bar{x}_n \to x_0 \\ x_n \neq \bar{x}_n}} \frac{f(x_n) - f(\bar{x}_n)}{x_n - \bar{x}_n}$$

determines the "continuous derivability" of the function f at the value x_0. In 1965, Paul Lorenzen (1915–94) called this concept, which was introduced by Guiseppe Peano (1858–1932) in 1892, "free derivability".

Cauchy's Concept of the Integral

In the "Algebraic Analysis", the integral was introduced as the inverse operation of differentiation, i.e. as the "indefinite integral"—and, consequently, as a "function".

Cauchy ends this tradition by introducing the "definite integral"—which is to say the integral as a "value".

We have already studied Leibniz' method of calculation of the area when one side is curved (pp. 29f) and seen that this method of Leibniz has only been known since the late twentieth century. Subsequently, if Cauchy proceeds in a similar manner, he could not have been influenced by Leibniz (or one of his followers).

This similarity is due to the visualization of the geometrical construction, not to the concepts.

- Where Leibniz takes a "curved line", Cauchy deals with a "function" $f(x)$ of an independent variable x.
- Where Leibniz introduces "points of division" B_k on the horizontal below, Cauchy sets forth "values of division" x_k in the assumed interval between x_0 and X.
- Where Leibniz uses "straight lines" D_k B_k, Cauchy relies on "values of a function" $f(x_k)$.
- Both take different quantities for their approximation:
    - Leibniz takes an "area" which *nearly fits*, namely the rectangles B_k P_k.
    - Cauchy decides to choose the suitable approximation $(x_{k+1} - x_k) \cdot f(x_k)$, which is a "value".

- And of course, where Leibniz *proves* that his approximate area differs from the true area by a quantity which can be made smaller than any given quantity, Cauchy *defines* the analytical concept of the "definite integral" as a "limit":

$$\int_{x_0}^{X} f(x)\, dx = \lim_{n \to \infty} \sum_{k=0}^{n-1} (x_{k+1} - x_k) \cdot f(x_k).$$

Thereby Cauchy assumes $f(x)$ to be a "continuous" function. (The subscript "$n \to \infty$" is of course missing in Cauchy.)

To define something is an easy exercise. The real difficulty is to prove that the definition *makes sense*. In our case, "making sense" is to show that the defined "limit" is well-defined and unique—in other words, that this limit (a value) does not depend on the method of calculation, that there are many ways to arrive at that limit or, to be precise, that each division $x_0, x_1, x_2, \ldots x_n = X$ of the interval from $x = x_0$ to $x = X$ produces the same value.

In other words, there are two different approaches with their own intricacies. Whereas Leibniz had to deal with the estimation of the greatest possible error,

Cauchy is concerned with comparing those limits that result from different divisions of the interval and subsequently has to prove that they are all equal.

We may skip the details here, for they can be looked up in any modern textbook of analysis. We shall only examine the *structure* of Cauchy's argumentation.[3]

Cauchy's Basic Idea in His Proof of the Existence of the Definite Integral

The decisive step of Cauchy's proof is the validity of this equation:

$$\sum_{k=0}^{n-1}(x_{k+1} - x_k) \cdot f(x_k) = (X - x_0) \cdot f(x_0 + \theta \cdot (X - x_0)) .$$

- On the left, we have the sum of products $(x_{k+1} - x_k) \cdot f(x_k)$. Geometrically, these are the "steps" *below* the graph of the function, in case of "increasing" functions as those by Leibniz.
- On the right, we have just one product, consisting of the length of the whole interval and the value of the function at the value $x = x_0 + \theta \cdot (X - x_0)$, where θ is an unknown value between 0 and 1. For such a θ, $x = x_0 + \theta \cdot (X - x_0)$ is *any* value inside the interval from x_0 to X. —That this is possible, i.e. that a value with this property really exists, is guaranteed by the so-called (and very famous) *"Intermediate Value Theorem"*. Bolzano formulated this theorem very precisely and proved it in 1817. We did not touch on this part of Bolzano's treatise. The decisive point is that the Intermediate Value Theorem is only valid for "continuous" functions. That is why Cauchy demands the continuity of $f(x)$ in his definition of the definite integral $\int_{x_0}^{X} f(x)\, dx$.

Let us summarize. A (*finite*) sum of products $(x_{k+1} - x_k) \cdot f(x_k)$ may be replaced by a single product which is made from the length of the interval and some value of the function inside of the interval.

The problem with this proof is the comparison of *different* divisions of the interval such as $x_0, x_1, x_2, \ldots x_n = X$ and $x_0, x_1', x_2', \ldots x_m' = X$. The idea is to "refine" each of these divisions in such a way that both refinements coincide. This is not hard but a little bit technical.

For each part of such a common "refinement"—say $x_0, x_1'', x_2'', \ldots x_l'' = X$—we have the above equation and in this way is the sum of products with x_i''-intervals transformed to a single product of length $(x_{k+1} - x_k)$.

The rest goes without saying.

---

[3] It is shown in *Die Analysis im Wandel und im Widerstreit*, pp. 339–343.

One cannot deny that Cauchy's proof of that useful equation which transforms a sum of products into a single product (with the length of the interval as one factor) is remarkable. To do so, Cauchy relies on a very general theorem regarding the estimation of quantities which he had already stated in his first textbook of analysis. (His textbook on the calculus of integrals was printed only eight years later, in 1829.)

When adapted to a form which is required for the proof of the existence of the definite integral, this theorem reads:

> Theorem. *Let b, b', b'' ... denote n quantities of the same sign and a, a', a'', ... be the same number of arbitrary quantities, then we have*
>
> $$\alpha b + \alpha'b' + \alpha''b'' + \ldots = (\alpha + \alpha' + \alpha'' + \ldots) \cdot M(b, b', b'', \ldots),$$
>
> *where $M(b, b', b'', \ldots)$ denotes any quantity in between the greatest and the smallest of the b, b', b'', ...*

It is very clear that "quantity" is a forceful basic concept in Cauchy 's analysis.

What Is Cauchy's "x"?

A final summary:

> Cauchy splits the formerly *unique* "x" of the old "Calculus of Expressions" for his new "Calculus of Values" into *two*:
>
> 1. a *small* "x" *without index* to denote the *"independent" variable,*
> 2. a *small and indexed* "x" or a *capital (and* upright*)* "X" to denote a *"value"* of the independent variable.
>
> In "determining precisely" these conditions he makes "all uncertainty disappear".

Literature

Cauchy, A.-L. (1821). *Cours d'Analyse*. In *Œuvres Complètes*, vol. 3 of series II. Paris: Gauthier-Villars, 1897. http://gallica.bnf.fr/ark:/12148/bpt6k90195m.r=cauchy+oeuvres.langFR. German: Itzigsohn 1885.

Cauchy, A.-L. (1823). *Resumé des Leçons données à l'École Royale Polytechnique sur le Calcul Infinitésimal*. In *Œuvres Complètes*, vol. 4 of series II, pp. 5–261. Paris: Gauthier-Villars, 1899. http://gallica.bnf.fr/ark:/12148/bpt6k90196z.r=cauchy+oeuvres.langFR.

Cauchy, A.-L. (1829). *Leçons sur le Calcul Différentiel*. In *Œuvres Complètes*, vol. 4 of series II, pp. 265–572. Paris: Gauthier-Villars, 1899. http://gallica.bnf.fr/ark:/12148/bpt6k90196z.r=cauchy+oeuvres.langFR.

Cauchy, A.-L. (1833). *Résumés analytiques (Turin)*. In *Œuvres Complètes*, vol. 10 of series II, pp. 7–184. Paris: Gauthier-Villars, 1895. http://gallica.bnf.fr/ark:/12148/bpt6k902022.r=cauchy+oeuvres.langFR.

Cauchy, A.-L. (1837). Extrait d'une lettre à M. Coriolis. In *Œuvres Complètes*, vol. 4 of series I, pp. 38–42. Paris: Gauthier-Villars, 1884. http://gallica.bnf.fr/ark:/12148/bpt6k90184z.r=cauchy+oeuvres.langFR.

Cauchy, A.-L. (1849). Sur quelques définitions généralement adoptées en Arithmétique et en Algèbre. In *Œuvres Complètes*, vol. 14 of series II, pp. 215–226. Paris: Gauthier-Villars, 1938. http://gallica.bnf.fr/ark:/12148/bpt6k90206f.r=cauchy+oeuvres.langFR.

Itzigsohn, C. (1885). *Algebraische Analysis von Augustin Louis Cauchy*. Berlin: Springer. http://gdz.sub.uni-goettingen.de/dms/load/img/?PPN=PPN379794896.

Lakatos, I. (1966). Cauchy and the Continuum: The Significance of Non-standard Analysis for the History and Philosophy of Mathematics. Cited from Lakatos 1980, vol. 2, pp. 43–60.

Lakatos, I. (1980). *Philosophcal Papers* (2 vols) vol. 1: *Mathematics, science and epistemology*. Cambridge University Press.

Laugwitz, D. (1987). Infinitely small quantities in Cauchy's textbooks. *Historia Mathematica, 14*, 258–274.

Lorenzen, P. (1965). *Differential und Integral. Eine konstruktive Einführung in die klassische Analysis*. Frankfurt am Main: Akademische Verlagsgesellschaft.

Mayer, J. T. (1818). *Vollständiger Lehrbegriff der höhern Analysis*. Göttingen: Vandenhoek und Ruprecht. urn:nbn:de:bvb:12-bsb10082386-8.

Peano, G. (1892). Sur la définition de la dérivée. *Mathesis, 2*(2), 12–14.

Riede, H. (1994). *Die Einführung des Ableitungsbegriffs*. Mannheim, Leipzig, Wien, München: BI Wissenschaftsverlag.

Robinson, A. (1963). *Introduction to model theory and to the metamathematics of algebra*. Amsterdam: North-Holland Publishing Company.

Spalt, D. D. (1981). *Vom Mythos der Mathematischen Vernunft*. Darmstadt: Wissenschaftliche Buchgesellschaft, [2]1987.

Spalt, D. D. (1996). *Die Vernunft im Cauchy-Mythos*. Thun und Frankfurt am Main: Harri Deutsch.

Spalt, D. D. (2002). Cauchys Kontinuum: Eine historiografische Annäherung an Cauchys Summensatz. *Archive for history of exact sciences, 56*, 285–338.

Spalt, D. D. (2015). *Die Analysis im Wandel und im Widerstreit*. Freiburg: Karl Alber.

Chapter 11
The Interregnum: Analysis on Swampy Ground

On the Utility of History of Mathematics

The Special Quality of Our Perspective

The last chapter explained Cauchy's basic view of analysis. This explication claims that it would meet Cauchy's approval in case he were to read it as our contemporary (whatever this may mean).

Moreover, this explication is nonetheless one by Cauchy himself—or one which he would have been able to give. This exposition of Cauchy's basic view of analysis is an exposition *from today,* from our actual viewpoint—what does this mean?

The given exegesis of Cauchy's basic view of analysis is not a *translation* of his words into modern language. (For otherwise, we would not have spoken of "quantities", as this notion does not exist at all in modern analysis.) Instead, our aim is that:

1. The focus of the previous chapter was Cauchy's basic view of analysis *within the context of his own concepts.* (Other chapters present the concepts of other mathematicians.)
2. This was *opposed* to other basic views of analysis: (a) the "Algebraic Analysis" of Euler and Lagrange, (b) the actual (standard) analysis, and (c) the actual Nonstandard-analysis.
3. The last two views were beyond Cauchy's grasp, for they did not exist in his times. Consequently, these views are *precisely* the *addition* which history of mathematics can utilize in order to *assess the potential* of Cauchy's analysis retrospectively.
4. These opposing perspectives *supplement* the theory created by Cauchy. They show aspects that *were not available* to Cauchy (as well as to his contemporaries). Yet, they *do not change* Cauchy's mathematical theory.

© The Author(s), under exclusive license to Springer Nature Switzerland AG 2022   149
D. D. Spalt, *A Brief History of Analysis,*
https://doi.org/10.1007/978-3-031-00650-0_11

5. These are *eminently mathematical* aspects. They allow for *definite* answers to questions of the following kind: (a) Is the "characteristic function" of the rational numbers (i.e. $f(x) = 1$ if x is rational and $f(x) = 0$ otherwise) in Cauchy's view a "function"? (b) What does Cauchy understand by "limit": a *uniquely* determined value or *several* values? (c) Does Cauchy deal (in modern language) with *usual* "convergence", with "uniform" convergence, or with "continuous" convergence or with something quite different? (d) Is Cauchy's notion of "differentiability" our modern "continuous differentiability"? (e) Is his "Sum Theorem" true or false? etc.

6. This is the very technique of obtaining a *historical as well as philosophical assessment* of the *mindset* of the mathematician under discussion. This allows for the building of bridges to general history as well as to the history of philosophy.

What Additional Knowledge Have We Gained That Was Not Available to Cauchy and to His Contemporaries?

Surely, Cauchy was very aware of the dissenting character of his lectures from the "Algebraic Analysis" of Euler and Lagrange. He explicitly declared this in the preface to his first textbook (p. 116). Retrospectively we can say that such a fundamental transformation remains unique within the history of analysis.

However, from today's point of view, we can give a more detailed account:

> Cauchy presented his very own view of analysis. It differs mathematically very clearly from all current versions of analysis, especially from standard-analysis and Nonstandard-analysis.

Crucially, we have to add:

> Such an *understanding* was *principally impossible* for Cauchy and for his contemporaries—for the conceptual framework of these versions did not exist at their time.

In the year of publication of Cauchy's first textbook, 1821, it was ABSOLUTELY NOT IMAGINABLE that there may exist DIFFERENT (and consequently, competing) *mathematical theories about* ONE *topic.* The most prolific mathematicians of that time were just engaged in creating a new *geometry,* a fact that was going to shock the scientific community. Thus, it still took many decades until "non-Euclidean" geometry was accepted as a *decent mathematical theory* (besides the earlier one which only then was called "Euclidean")—as a different but *equally exact* and *mathematically legitimate* theory.

However, geometry is no exception and the same can be said about analysis. Only the twentieth century brought this awareness of the essential ways in which mathematics develops and changes, and even today this knowledge is still very far from being widespread.

The Teaching of Analysis Without a Curriculum

The higher schools had already started to teach analysis in the eighteenth century. In Germany, the successful textbooks of Abraham Gotthelf Kästner (1719–1800), written also for private studies, illustrate this. In the nineteenth century, this tendency increased enormously. The beginning of the era of industrialization required ever more mathematical skills for ever more people.

Which kind of analysis was taught at the universities?

Seen from today's point of view, one could ask more precisely: did they teach "Algebraic Analysis" or the "Analysis of Values"?

However, this detailed question does not make sense—at least if one were expecting a *considered, meaningful* answer from the point of view of those past teachers. As we have just stated: save Cauchy, *surely* no mathematician of the first half of the nineteenth century was aware that *different theories of analysis* were possible. Of course, Cauchy knew only one rival theory: the "Algebraic Analysis" of Euler and Lagrange. (This is, strictly speaking, still true today.)

Instead, analysis then was simply analysis (and as a rule, so it remains today): a theory thought (or supposed) *to be undivided.* It was *this* doctrine which was taught as "calculus" to the students.

For example, in my experience and to my knowledge, it had not been realized until 1991 that the *very* meaning of the concept of "convergence" depends on the *precise* meaning of the concept of the "value of a function" (see pp. 135f).

For a very long time, the *concepts* of analysis—like *all* mathematical notions—were taken to be *clear-cut* and *unchangeable*. As objects of a "Platonic world", they seemed to be eternal.

The *insight* that this is false—that the non-existence of an eternal *code of mathematics,* for mathematics is created by a *specific* human culture, *like all the other sciences*—is a new one. This is far from common knowledge. (Although, even today nobody is really able to conjecture *where* this super code of mathematics could be written down.)

In case you know only *one* version of analysis, there is no basic problem with the curriculum. Then the only maxim is teach *the* analysis, particularly those parts which are *important* for your students!

This is what was usually done. Each university or college developed a curriculum of its own, depending on the knowledge of the local staff. There was no awareness of the necessity of synchronizing teaching in different places (in order to *compare* students' ability or, what amounts to the same, *to keep the unity of mathematics*).

> In the early nineteenth century, nobody realized that, as a consequence of this missing coordination of teaching, the unity of analysis (to be understood as a system without contradictions) was at stake.

Actually, not one of the scientists of that time was *able* to realize this—this is an objective statement. The standard of knowledge had not yet reached the level to engender such a *possibility.* Only today have we arrived at this insight. Only today we know:

> If the *basic concepts* of a mathematical theory are not stated with *absolute clarity,* the subsequent mathematical propositions become somewhat *mathematically imprecise.*
> Here, *mathematically imprecise* entails nothing other than that the relevant propositions might produce a contradiction, be it in respect to each other or be it in respect of the propositions of other authors.
> The "contradictions" under consideration are those which could, in principle, be detected by the mathematicians of that time. They are not "contradictions" which could only be detected with the benefit of hindsight.

We have already mentioned two such contradictions, namely *controversial* theorems: the "Fundamental Theorem of Functions" (p. 128) and "Cauchy's Sum Theorem" (p. 138).

Analysis as Freestyle Wrestling

Indeed, analysis developed in the nineteenth century in the manner of freestyle wrestling or in today's world, like mixed martial arts. This will be shown in the current chapter with reference to the following three points:

A. Missing conceptual precision
B. Political rather than rational argumentation
C. Parallel methods which differ in precision

A. Missing Conceptual Precision—or—Riemann Invented the Modern Concept of Function

Already in the preface to Cauchy's first textbook of analysis—as we propose to call it: "Analysis of Values"—we can read: "In determining *the values* of the quantities I will make all uncertainty disappear." (p. 116).

As "function" has been the central concept of analysis since its invention by Euler, the very first consequence of Cauchy's dictum is this: *determine the concept "value of the function" precisely!*

Cauchy did so—we discussed this at length (see pp. 124f. (Some may have thought: much too long!)

And what about Cauchy's colleagues and successors?

The astonishing answer is: for a whole generation, or fully 30 years, *nothing* happened. This *nothing* has to be taken seriously:

> Neither was Cauchy's definition of "value of the function" accepted by his colleagues or successors, nor was it opposed by a competing one.

The different positions are briefly given as examples below:

1. Dirichlet in 1837 Carl Gustav Jacob Jacobi (1804–51) sings Dirichlet's praises. In a letter to Alexander Humboldt, Jacobi writes gushingly:

> Only he, not me, not Cauchy, not Gauß, knows what a completely strict mathematical proof is. If Gauß says he has *proved* something, I take it as likely; if Cauchy says so, there are equally many pros and cons to bet; if Dirichlet says it, it is *sure*.

Let's see!

Dirichlet is generally taken to be the inventor of the modern concept of a function. This is incorrect. I do not know *one single* place where Dirichlet defined the concept of "value of the function".

Even worse, according to Dirichlet, the "value of the function" is not necessarily *uniquely* determined—as is demanded in all actual textbooks of analysis. On the contrary, Dirichlet invented a notation, which is still used today, for functions that have two different limits on either side, depending on whether the left-hand side or the right-hand side is taken (for Dirichlet, "−0" or "+0"). He writes the following:

> If the function $f(x)$ undergoes a sudden change for singular values of x without becoming infinite then there, then this curve consists of several interrupted parts. For every such abscissa, we detect two ordinates, actually, one of them belonging to the ending and the other one to the starting piece of the curve.
>
> In the following, it will be useful to distinguish those two values of the function and denote them by $f(x - 0)$ and $f(x + 0)$.

That is to say, if the function has a jump for the value x, Dirichlet distinguishes *explicitly* the TWO "values of the function" to be $f(x + 0)$ and $f(x - 0)$.

Not only this! Some pages later, Dirichlet treats an incidence of *just the same kind*—but his calculations lead to a *third* "value of the function":

> When a disruption of the continuity occurs and, consequently, the function $f(x)$ has two values, the [trigonometric] series [of this function], which naturally has only one value for each x, gives half of the sum of these values.

That is, the function has at this point *two* "values" (in some sense determined and thereby following Cauchy) $f(x + 0)$ and $f(x - 0)$—as well as a *third* "value" which is algebraically determined from the trigonometric series to be the mean of the other two.

A completely strict mathematician is *expected* at this point to declare which one (or more) of these "values" is (are) to be taken as the "value(s) of the function". However, Dirichlet keeps completely silent on this. This is not strict at all!

Interestingly, there is one formulation by Dirichlet from which we can understand that he, *as well as Cauchy* before him, took the "Fundamental Theorem of Functions" to be *self-evidently* true.

We have just read that Dirichlet did *not accept* Cauchy's concept of "value of a function" which is (not strictly logical, but based on the facts) a *necessary* assumption of Cauchy's Fundamental Theorem of Functions. In other words, if the following interpretation were true, we would have to conclude that Dirichlet's thinking regarding the basic notions of analysis was self-contradictory.

By virtue of lecture notes from the year 1854, we know that Dirichlet defined "continuity" as follows:

> $y = f(x)$ is called a *continuous* and *definite* or *one-valued* function of x, if each value of x only belongs to *one* value of y and if each gradual change of x corresponds to a gradual change of y, i.e. if for some fixed x, the difference
>
> $$f(x + h) - f(x)$$
>
> for steadily decreasing h converges on the limit zero.

I confine myself to the following simple interpretation, which goes as follows:

> (i) In his first sentence, Dirichlet gives *one single* definition. (ii) In this definition, Dirichlet uses *two* names: "continuous" and "definite or one-valued". (iii) Consequently, Dirichlet took *both* names *to mean the same*. (iv) The meaning of "definite or one-valued" is self-explanatory. (v) To sum up, in this definition, Dirichlet supposes the validity of the "Fundamental Theorem of Functions"—end of the simple interpretation.

The rules of higher rhetoric allow for a *contradictory* interpretation of Dirichlet's passage. We shall leave the details to the specialists of rhetoric and plainly state its result: it claims that Dirichlet is giving *two* definitions in his first *single* sentence. However, the question arises: is it legitimate that Dirichlet opens his lecture by giving such an artificial and pretentious formulation, wherein he defines *two* different concepts in *one single* sentence? In one sentence, which allows a *simple* understanding with a fundamentally different meaning of the two given notions? For indeed, this sentence is the very first of Dirichlet's lecture notes.

I believe this to be unrealistic and therefore keep to my straightforward interpretation.

This simple interpretation fits well to the following formulation from (p. 131) of the lecture notes (*emphases* original): this function is

> *thoroughly continuous*, i.e. *everywhere definite and specified, finite and changes gradually.*

To Dirichlet as to Cauchy, the "and changes gradually" is a *consequence* of the "definite and specified", whereby "definite" implies at least "finite".

2. Riemann in 1851 Dirichlet's pupil Bernhard Riemann conceived things differently. The first half of the first sentence of his doctoral thesis, published in 1851, goes as follows:

> If one thinks z to be a changing quantity, which gradually can take all possible real values and if to each of its values there corresponds one singular value of an indefinite quantity w, then w is called a function of x ...

Herein Riemann states point-blank: all you need in order to have a "(real) function" is, that to each "value" exists a "corresponding" *single* "value"—neither more nor less. *How* this "correspondence" is accomplished (be it through a "calculatory expression" as with Euler or a "dependence" as with Cauchy or otherwise) is left completely open by Riemann. He says no more than "(definite) correspondence".

That sounds as if Riemann simply drew the consequences from Cauchy's definition of the "derivative"—which, ironically, does not fit at all Cauchy's own concept of function (see p. 144). Riemann was going to change all this, because by now the "Calculus [of differentials and integrals]" *required* the *function* "derivative". Consequently, the concept of function needed urgently to be *adjusted* to fit new developments.

> Riemann coins the concept of "function" which is still valid today: to each "value", *x uniquely* corresponds a "value of the function" $f(x)$—nothing more, plainly this. Nothing is assumed about the manner or the realization of this correspondence.

Thus the *matter* is stated. Obviously, Riemann's *style* differs from ours. It is his *style* to talk about "quantity" when we, today, mostly speak about "sets" and "elements", maybe even about "ordered sets". However, these are differences of expression, not of mathematics. The mathematical objects "function" and "value of the function" of Riemann and of ours are the *same*.

Clearly, concerning philosophical aspects, this is put somewhat roughly: *style* and *content* cannot be separated completely. However, this is not the place to differentiate that precisely; our formulation shall suffice here.

As one can easily imagine, in spite of Riemann's exceptional mathematical genius, the publication of his doctoral thesis was not an instant success. It did not lead to the immediate acceptance of Riemann's new view in 1851. Initially, many mathematicians did not follow Riemann, among them his illustrious antagonist Weierstraß—at least almost until the very end of his career. Certainly, we shall return to this.

In the long run, it turned out that Riemann was on the right track, especially when he opted for the most general concepts possible for both a function and a value of the function. Bolzano had already anticipated those (p. 111), but as this revolutionary thinker had been relegated from university by the conservative government and banned from publishing, his idea remained unpublished in the nineteenth century. Riemann can therefore take all the credit for having invented the two most central concepts of actual mathematics and for their introduction into the mathematical discourse.

3. Björling in 1864 Anybody who did not live in Paris in the midst of the nineteenth century and who did not have an excellent teacher but *really* wanted to study analysis sooner or later realized that he or she had to study Cauchy's textbooks. All those who *only* had Cauchy's books at their disposal might have been well motivated to read them *very carefully*.

Such a studious student was Emanuel G. Björling (1808–72) from Sweden. The relevant lexicons do not know him. A footnote of one of his treatises, published in 1854, reads as follows:

> We should record that it is said for a purpose and not because of some laxity that a function of x which changes continuously in between its two limits is only allowed to have a single, unique, finite value for each limit of x.

A careful inspection shows that this is exactly what the "Fundamental Theorem of Functions" says! (For Björling demands in no uncertain terms: *if* the value of the function for some "limit"—or a "value"—is *unique*, *then* the function is *"continuous"* at this value.)

I have been unable to detect any other *explicit* statement by authors other than Björling and Cauchy, which *clearly* embraces the content of the "Fundamental Theorem of Functions" (and which therefore *presumably* assumes Cauchy's concept of the value of a function). The *possible* exception of Dirichlet has already been discussed.

B. Political Instead of Rational Reasoning

As already mentioned on pp. 133f, Cauchy's (only today so-called) "Sum Theorem", stated in his textbook on analysis, was repeatedly criticized. The most popular criticisms are those by Niels Henrik Abel from 1826 and by Philipp Ludwig Seidel from 1850. *How do these criticisms run?*

The answers to this question are as sobering as they are interesting:

> Neither Abel nor Seidel criticize Cauchy's proof.
> Instead, they just state their *different* opinion and try to justify *it*.
> Both critics do not criticize but proclaim yet *another point of view* on the subject.

Clearly, this is a *political* way of reasoning, not a mathematical one.

Just as in our everyday experience, political debates run as follows: the specific views are presented and supported by some arguments, and all opposing views are condemned—*just because* they contradict the speaker's own views, but not because they are *inconsistent in themselves* or *unconvincing* (or even *illogical*). This seems to be the ubiquitous practice of political debates today.

Abel and Seidel conceived analysis just *in contrast* to Cauchy. That is not objectionable. At the same time, this does not justify the judgement that Cauchy's theorem is *wrong*. The only thing Abel and Seidel are *justified* in saying is: "We see things differently from Cauchy!" This would be totally legitimate, and if they were to produce their arguments, it could even be constructive.

1. Abel did not give any argument. He simply claimed Cauchy's theorem to "suffer from exceptions" (a polite formulation towards the authority of Cauchy). He even produced an *example* of such an alleged "exception". But he proved neither (a) that *and why* Cauchy's argumentation should be wrong, nor (b) that and why his example should contradict Cauchy's theorem. Abel argued by brute force: "I differ from Cauchy and in this way!" No argument whatsoever, neither *against* Cauchy nor *in favour of* the correctness of his own claim.

> The mathematical value of this Abelian *claim* is obviously nil.

Thus, it is very strange that the widespread assessment of Abel's criticism of Cauchy was completely different. Practically all mathematicians (and historians of mathematics) of today praise Abel to be a perspicacious critic of Cauchy. On what grounds?

A possible answer might be this: the proponents of this assessment did not familiarize themselves with Cauchy's way of thinking or with that of Abel (which in this case is less important). (What would have to be done in this case is explained on pp. 140f.) They do not rate the mathematical stringency of the given arguments by Cauchy and Abel. Instead, they simply ask "who is right?" Just as if there were *only one single* CORRECT *construction of analysis.* As stated before, this is a *purely political* (and one-sided) judgement.

However, this stage of mathematical development has been overcome for at least 50 years. (If we include the intuitionists, even for a century.) Nevertheless, the majority of mathematicians (and worse, of the historians of mathematics, too) are not willing to take notice of that. Why not?

2. Seidel, a generation later, was much more constructive than Abel. Indeed, Seidel, too, did not try to detect Cauchy's reasoning error. (As we know, he would not have succeeded as such an error *does not exist.*) Still, Seidel at least *justified* his view—thereby *showing* his dissenting conception from Cauchy's analysis in strict and clear mathematical terms.

This does not mean that Seidel *specified* this *dissent*—not at all! Instead, Seidel produced *clear arguments* for his own view, and this we are going to examine here. (It is the job of the historian to *compare* this way of argumentation with Cauchy's.) It was Seidel's achievement to coin the concept which is called "uniform convergence" today. (To be precise, Seidel came up with the *opposite* of the concept of "uniform convergence" and called it "arbitrarily slow convergence". But clearly, the definition of a concept also defines its opposite.)

Seidel offered a *precise* definition of his new concept and with this notion he formulated a theorem that logically contradicts Cauchy's Sum Theorem—*if* Cauchy's concepts (and especially his concept of "convergence" and, consequently, his theorem) are taken *in a way which was not intended* by Cauchy, that is to say: as understood and taken by Seidel

For those who want to know more: Seidel formulated and proved the following theorem:

> Theorem. *If you have a converging series which represents a discontinuous function and if the single terms of the series are continuous functions, than it must be possible to specify values in the immediate neighbourhood of the place at which the function jumps, for which the series converges arbitrarily slowly.*

Seidel's theorem says: if a "converging" series of continuous functions represents a function which is discontinuous at the value $x = x_0$, then the convergence of the series is not the usual one but one which is only "arbitrarily slow".

Let us consider it the other way round: if we *only* demand the usual convergence of the series at the value $x = x_0$ (as was done by Cauchy, in Seidel's understanding, which as we already know is erroneous), then the continuity of the function at the value $x = x_0$ may not be guaranteed.

Seidel has a complicated way of constructing his analysis. The same holds for today's standard-analysis.

One more final word about the invention of the concept of "uniform" convergence for this notion is somewhat important in modern analysis.

As stated above, Seidel invented this concept in 1850 (more precisely, its opposite). However, he was not the only one to do this. Abel published a treatise where he used the name "steadily convergent" ("beständig konvergent") already in 1829, and in a letter published in 1830, Abel explains his understanding of this notion. Christoph Gudermann (1798–1852) *presumably* took this concept and later taught it to his gifted student Karl Weierstraß. It was Weierstraß who then made this concept well-known (under the name "uniform" convergence, "gleichmäßige" Konvergenz)—we shall return to this later.

Parallel Methods with Different Levels of Precision

It is central for Cauchy to use the concept of "limit", written "lim". This notion is the *technical realization* of the programme presented in the preface to his first textbook: the *precise determination* of the relevant "values" of the "variable". The "limit" mediates between the "variable" and its "values".

$$\text{"For} \quad x = 0, \quad \text{we have} \quad \lim r_n(x) = 0\text{"}$$

means the "variable" $r_n(x)$ (with the two variables n and x) has for the "value" $x = 0$ and $n \to \infty$ the *unique* "value" 0.

Unlike us today, Cauchy uses the operator "lim" always *without subscripts:* " $\lim_{\dots \to \dots}$ ". This can be *explained* by the fact that Cauchy always means *all* possible limits—at least in the context of his accompanying text. Thus, Cauchy has *all* limits to make up the "value(s) of the function" for the "value" $x = X$, and therefore any specification of the limits is completely *superfluous*. With the above,

$$\text{"}\lim r_n(x) = 0 \quad \text{for} \quad x = X\text{"} \qquad \text{Cauchy means our} \qquad \text{"} \lim_{\substack{n \to \infty \\ x_k \to X}} r_n(x_k) = 0 \text{"}.$$

All the limits are taken, $n \to \infty$ as well as $x_k \to X$ and all their inter-dependencies whatsoever.

Of course, if one uses *another* concept of "lim"—as, e.g., Riemann does—one *must* specify exactly *which* limits are meant. In this case, one *must* use subscripts with "lim". We today have learnt analysis in another way from that developed by Cauchy. That is why we automatically tend to read Cauchy's formula in the following way:

$$\text{"}\lim r_n(x) = 0 \quad \text{for} \quad x = X\text{"} \qquad \text{means} \qquad \text{"} \lim_{n \to \infty} r_n(X) = 0 \text{"}.$$

We take *only one* limit: $n \to \infty$; the other one: $x_k \to X$ is *omitted,* and also the possible dependencies between them. All this because today we replace Cauchy's concept of "value of the function" by that of Riemann. *Cauchy understood " lim" differently from us today.*

This leads to the modern misunderstanding of Cauchy, and this misunderstanding transforms (in modern language) Cauchy's concept of "continuous convergence" to the *simple* "convergence".

Not every mathematician since Cauchy has used the concept of "limit". For instance, Seidel's treatise from 1850, mentioned above, contains many considerations of limits, but the operator " lim" is never used. For example, his comments on the formula $**$ on p 139 go as follows:

$$s_n(x + \alpha) - s_x(x) < \varepsilon_1$$

$$r_n(x) < \varepsilon_2$$

$$r_n(x + \alpha) < \varepsilon_3$$

where ε_1, ε_2, ε_3 designate arbitrarily small absolute quantities and all the inequalities are taken independently of the positive or negative sign.

ε_1, ε_2, ε_3 are meant to designate (small) numbers. (Actually, Seidel used other Greek letters, but this does not matter mathematically. The use of epsilons facilitates our understanding as we are used to them.) By the way, Seidel understands by "x" a "value", not a "variable".

Today the method used by Seidel is called **"epsilontics",** for today the Greek letter epsilon usually denotes certain *very small* tolerances. If it is possible to show—let us stick to the above example—that we have

$$s(x + \alpha) - s(x) < \varepsilon_1 + \varepsilon_2 + \varepsilon_3$$

as long as we have $\alpha < \delta$, it is proved that we actually have

$$\lim r_n(x) = 0.$$

(Of course, the problem is always somehow to specify the security margin δ.) Then it is *proved:* $r_n(x)$ can be made *smaller than* any arbitrarily small given number $\varepsilon_1 + \varepsilon_2 + \varepsilon_3$ (independent of the sign). Or the other way round: it is *impossible* that $\lim |r_n(x)| = h$ with a number $h > 0$—consequently, only $\lim r_n(x) = 0$ is left.

The usage of the operator lim is often called **"limes calculation".** The name is somewhat inappropriate, for we do not calculate *with* the " lim", but we calculate *the* limit. However, we cannot change the common terminology.

Using both names, we can summarize:

| Analysis is possible with epsilontics as well as with the help of limes calculation. |

The difference between the two methods is that:

| The limes calculation needs a variable "quantity", epsilontics does not. |

That is why epsilontics is more general than limes calculation.

The general implementation of limes calculation is due to Cauchy. He used the operator lim as a central tool in realizing his programme, for this operator *mediates* between the "variable" (which is a "quantity") and its "values".

In contrast to Cauchy's analysis, modern theory lacks the notion of "quantity" and certainly does not have it as a foundational concept. Therefore, if modern analysis operates with limits, *another understanding* of the operator lim than that of Cauchy is required. One *expression* of this other understanding is today's *demand* clearly to specify the limes in the subscript " $\lim_{...\to...}$ ".

The Foundational Uncertainty of Limes Calculation: A Prominent Misunderstanding by Prominent Scholars

Today's mathematicians are well versed in epsilontics. This may lead to a disregard of earlier limes calculations. For example, the distinguished mathematician of the later twentieth century, Detlef Laugwitz, claimed, in all seriousness, that Riemann's and Cauchy's formulations of the basic definition of convergence were wrong:

> Even more surprisingly, one finds [in Riemann] obviously wrong formulations as in the case of the convergence criterion of a sequence $s_n(x)$ [...] Almost literally, the same misleading formulation can repeatedly be found in Cauchy since his *Cours d'analyse* of 1821.

This judgement, which is printed in a specialized textbook, is far from modest. However, is it well-founded?

A closer reading shows: no, this judgement is not substantiated. It is based on a misreading. This will briefly be explained, for it shows the fundamental problem with modern limes calculation.

At first, one easily realizes that this formulation which Laugwitz judges to be "obviously wrong" is to be found not only in Cauchy and Riemann but also in Georg Cantor and even in Karl Weierstraß, who is commonly judged to be extremely meticulous. This must create some doubt. So let us take a closer look.

Laugwitz criticizes some notes, taken from Riemann's lectures by one of Riemann's students. They state (*emphases* are original):

> An infinite series, the terms $u_0, u_1, \ldots$ of which follow a law of formation, is convergent if the sum
>
> $$s_n = \sum_{m=1}^{m=n} u_m$$
>
> for increasing n approaches a fixed limit which is called the value of the series. Therefore, the general condition for convergence is that
>
> $$s_{n+k} - s_n \, ,$$

for *arbitrary* k and increasing n, decreases without end, or in other words that for every k

$$\lim_{n=\infty} \sum_{n+1}^{n+k} u_m = 0.$$

Weierstraß writes:

> In case of such an unconditionally summable series, one easily proves that
>
> $$\lim_{n=\infty} |s_{n+k} - s_n| = 0$$
>
> for each k, i.e. the change of s_n becomes arbitrarily small for increasing n.

The problematic formulation is the condition for the segment length k. Riemann says "for *arbitrary* k", while Weierstraß writes "for each k". So what constitutes the problem?

The problem Laugwitz touched upon is best understood through the example of the "harmonic series":

$$1 + \tfrac{1}{2} + \tfrac{1}{3} + \tfrac{1}{4} + \tfrac{1}{5} + \cdots$$

On p. 85, we have already studied it in detail and have shown:

> However far we proceed in the series, there *always* exists a segment of length k, where the sum of its terms exceeds $\tfrac{1}{2}$:
>
> $$\tfrac{1}{n+1} + \tfrac{1}{n+2} + \ldots + \tfrac{1}{2n} > n \cdot \tfrac{1}{2n} = \tfrac{1}{2}.$$

Therefore, it does not converge. However, if you specify the length k of the segment already *before* you decide on the length N of the section, you get for the "harmonic series" the following:

> For each tolerance ε and for any length k for a segment, there exists a length N such that *for any* section of length greater N each partial sum of a segment of a length which is not larger than k is less than ε. Expressed algebraically,
>
> $$\tfrac{1}{n+1} + \tfrac{1}{n+2} + \ldots + \tfrac{1}{n+k} < \varepsilon,$$
>
> *if only $n > N$.*

That is to say, *at first* one decides on the value of k (and that of ε) and *only afterwards* does one determine the value of N. It is $N = N(k)$ (or, more exactly, $N = N(k, \varepsilon)$).

One can easily be convinced of it. The condition sounds like a generalization of the necessary condition for the (single) terms of a converging series to become arbitrarily small. In epsilontic formulation, for each tolerance ε, there exists a section of length N such that

$$\tfrac{1}{n} < \varepsilon \qquad \text{if} \quad n > N.$$

That is true, because one only has to choose $N > \tfrac{1}{\varepsilon}$ (which is always possible; why actually?); for then we have $\tfrac{1}{N} < \varepsilon$, and from $n > N$, we get $\tfrac{1}{n} < \tfrac{1}{N}$—all is done: $\tfrac{1}{n} < \tfrac{1}{N} < \varepsilon$.

Now, *one* term is just a sum with only a single summand. Let us now take a sum with two summands:

$$\tfrac{1}{n} + \tfrac{1}{n+1}.$$

With the help of the larger of them (which is the first one), we easily get an upper estimate:

$$\tfrac{1}{n} + \tfrac{1}{n+1} < \tfrac{2}{n}.$$

It is the same with a sum consisting of k summands:

$$\tfrac{1}{n} + \tfrac{1}{n+1} + \tfrac{1}{n+2} + \ldots + \tfrac{1}{n+(k-1)} < \tfrac{k}{n}.$$

From $n + k > n$, it follows $\tfrac{1}{n+k} < \tfrac{1}{n}$ and consequently

$$\tfrac{1}{n+1} + \tfrac{1}{n+2} + \tfrac{1}{n+3} + \ldots + \tfrac{1}{n+k} < \tfrac{k}{n}.$$

And now the trick used in the first step is: for a given tolerance ε, we choose $N > \tfrac{k}{\varepsilon}$ and consequently $\tfrac{1}{N} < \tfrac{\varepsilon}{k}$ or $\tfrac{k}{N} < \varepsilon$. Then from $n > N$, it follows $\tfrac{1}{n} < \tfrac{1}{N}$ as well as $\tfrac{k}{n} < \tfrac{k}{N}$ and finally

$$\tfrac{1}{n+1} + \tfrac{1}{n+2} + \tfrac{1}{n+3} + \ldots + \tfrac{1}{n+k} < \tfrac{k}{n} < \tfrac{k}{N} < \varepsilon.$$

That was to be proved. The decisive point is that at first one chooses the length k of the segment and the tolerance ε—only afterwards the section length N.

We have learnt two things: (i) both propositions in the boxes are valid for the "harmonic series". Against our initial impression, they do not contradict each other. (ii) The proposition in the second box does not guarantee the convergence of the series!

After finishing this excursion into epsilontics, we return to Riemann, Weierstraß, and the others and go back to limes calculation!

The last two boxes show: for the characterization of convergence the segment length k is of importance. For we have just shown that the "harmonic series" complies with the condition of the last box—*nevertheless* it does *not* converge, as the box before it demonstrates. The condition in the last box is too weak and does *not* guarantee convergence.

Now the difficult part, the *interpretation* of the texts. A possible condition for convergence given by limits is:

$$\lim (s_{n+k} - s_n) = 0 \quad \text{for each } k.$$

The segment of the series under examination in case of the "harmonic series" is

$$s_{n+k} - s_n = \tfrac{1}{n+1} + \tfrac{1}{n+2} + \ldots + \tfrac{1}{n+k}.$$

Now, what about k?

Riemann says (p. 160) that the equation $\lim (s_{n+k} - s_n) = 0$ is valid "for every k".

Laugwitz claims: in saying this, Riemann fails to notice the weakness of the condition in the last box. In more detail, Laugwitz says: (i) in his definition of convergence, Riemann *demands* only the condition of the last box. (ii) However, the "harmonic series" *fulfils* just this condition (this we just proved). (iii) Nevertheless, the "harmonic series" *does not* converge (this is shown the box before that). (iv) *Consequently,* Riemann's definition of "convergence" is wrong.

Anyhow, this is only an assertion of Laugwitz. It is indeed possible to understand Riemann this way: he demands that after having fixed the section length N, the partial sums of *every* segment length k stay less than the tolerance ε. That is,

---

For each tolerance ε, there exists a section length N such that *from* this section length N *onwards*, every partial sum of segment length k is less than ε. In formulae,

$$\tfrac{1}{n+1} + \tfrac{1}{n+2} + \ldots + \tfrac{1}{n+k} < \varepsilon,$$

if $n > N$ AND FOR EVERY k.

This is different from what the last box states, because N is now determined *independently* from k! N is *not* an $N(k)$.

If the clause "for every k" is taken in this way, i.e. if the tolerance ε is de facto prescribed *independently* of the length k of a segment, then the penultimate box shows that the "harmonic series" violates *this* condition. Consequently, the "harmonic series" is *not* a counter-example for the condition of convergence when understood *in this way*.

To sum up: if Riemann is read and understood in the way which is explained above, then *his definition of "convergence" is correct.* And the very same holds for Weierstraß' formulation.

The problem with the limes calculation is this: a formulation like "it is demanded that $\lim_{n \to \infty} (|s_{n+k} - s_n|$ for all $k) = 0$ " is impossible for purely *syntactic* reasons: an equation does not allow for text. Subsequently to deduce from this that "Riemann and Cauchy are unable to define the concept of 'convergence' correctly!" is inadmissible because it is not properly justified.

Let me conclude in the following way:

1. In the first half of the nineteenth century, both methods, limes calculation and epsilontics, were practised *side by side.*

2. With some difficult issues, it is hardly possible to offer an *unequivocal* formulation in limes notation. For example, Laugwitz misunderstood Cauchy's and Riemann's characterizations of convergence. (With this late historiographical text, Laugwitz did not live up to his excellent mathematical renown.)

3. This problem simply disappears if the limit is taken *as a matter of principle for all* variables—as it was practised by Cauchy.

What is "*x* " After Cauchy?

> Before Riemann's clear definition of the value of a function to be *uniquely* determined, there was an anarchy of significations. "*x*" could designate *everything:* a "variable" as well as a "value".

We observed this already in Bolzano: even he used "*x*" to designate a "variable" *as well as* any of its "values" (p. 108).

Only Cauchy made things clear and distinguished *even in notation* scrupulously between "variables" and their "values" (p. 147). Remarkably, this Cauchyan manner of notation was not taken up by anyone, as far as I know.

Even Riemann's decision to demand the value of the function to be uniquely determined did not end this anarchy of significations.

> Even after Riemann had demanded that the "value of the function" must be *uniquely* determined, the notation "*x*" could designate a "variable" *as well as* a precise "value" of it.
>
> In just the same way, "$f(x)$" could be chosen as the name of a function (i.e. a "variable") *as well as* for a "value of the function" (i.e. the "value" of the function at the "value" *x*).

Cauchy's code of notation would have blessed analysis with greater clarity, even after Riemann.

Literature

Arendt, G. (1904). *Gustav Lejeune Dirichlets Vorlesungen über die Lehre von den einfachen und mehrfachen bestimmten Integralen.* Braunschweig: Vieweg. reprint VDM-Verlag Dr. Müller, Saarbrücken, 2006.

Björling, E. G. (1854). Om det Cauchy'ska kriteriet på de fall, då functioner af en variabel låta utveckla sig i serie, fortgående efter de stigande digniteterna af variabeln. *Kongl. Vetenskaps-Akademiens Handlingar, För År 1852,* 165–228.

Dirichlet, G. L. (1837). Ueber die Darstellung ganz willkührlicher Funktionen durch Sinus- und Cosinusreihen. *Repertorium der Physik, I.*, 152–174. see also: Kronecker & Fuchs (Eds.) 1889, 1897, vol. 1, pp. 133–160.

Kronecker, L. (vol. 1) & Fuchs, L. (vol. 2) (Eds.) (1889, 1897). *G. Lejeune Dirichlet's Werke.* Berlin: Georg Reimer.

Laugwitz, D. (1996). *Bernhard Riemann 1826–1866. Wendepunkte in der Auffassung der Mathematik.* Basel: Birkhäuser Verlag.

Detlef Laugwitz 2008. *Bernhard Riemann 1826–1866. Turning points in the conception of mathematics.* Boston: Birkhäuser Verlag. English: Abe Shenitzer.

N. N. (1886). *Ausgewählte Kapitel aus der Funktionentheorie,* Vorlesung von Prof. Dr. Karl Weierstraß, Sommersemester 1886. In Siegmund-Schultze, R. (Ed.) *Teubner-Archiv zur Mathematik* (Vol. 9) Leipzig: Teubner, (1988).

Neuenschwander, E. (1987). Riemanns Vorlesungen zur Funktionentheorie, allgemeiner Teil. *Mathematisches Preprint No 1086,* Technische Hochschule Darmstadt.

Neuenschwander, E. (1996). Riemanns Einführung in die Funktionentheorie.. In *Abhandlungen der Wissenschaften in Göttingen, Mathematisch-physikalische Klasse* (Vol. 44) third series. Göttingen: Vandenhoeck & Ruprecht.

Pieper, H. (1987). *Briefwechsel zwischen Alexander von Humboldt und C. G. Jacob Jacobi.* Berlin (east): Akademie-Verlag.

Riemann, B. (1851). Grundlagen für eine allgemeine Theorie der Functionen einer veränderlichen complexen Grösse. in: Heinrich Weber unter Mitwirkung von Richard Dedekind (Ed.). *Bernhard Riemann, Gesammelte mathematische Werke.* reprint of the second edition from 1892 (pp. 3–48). New York: Dover Publications (1953).

Seidel, P. L. (1850). Note über eine Eigenschaft der Reihen, welche discontinuirliche Functionen darstellen. In *Abhandlungen der mathematisch-physikalischen Classe der königlichen bayerischen Akademie der Wissenschaften,* Jahre 1847–49 (Vol. 5, pp. 379–393). München: Weiß'sche Druckerei.

Chapter 12
Weierstraß: The Last Effort Towards a *Substancial* Analysis

A Famous Man

Karl Theodor Wilhelm Weierstraß (1815–97) invented analysis as a precise mathematical doctrine. So, at least, goes the story told to every German student of analysis nowadays. According to that myth it was Weierstraß who was the first to establish clear concepts in analysis and who first carried out strict proofs with the help of "epsilontics" developed by him. Only in this way could the vague ideas of his predecessors be expressed clearly.

As with any commonplace, this is nonsense. To be sure, Weierstraß left a lasting legacy in analysis, but he did not achieve the marvels ascribed to him. As the foregoing chapters have shown, he could not have achieved this. *All* basic notions of analysis—function, differential, integral, continuity, convergence—had already been known for a long time while little Weierstraß learned to add. All, apart from a single notion: the concept of number. Here, indeed, Weierstraß produced a marvel: he invented a concept which none of his students was able to grasp and, in addition, not one mathematician after him has been able to reinvent, since about 1870!—this gradually became clear after a lecture note from winter semester 1880/81 emerged unexpectedly in the mathematics library of Frankfurt university in summer 2016 (see pp. 172f).

Having set out to study the development of the basic concepts of analysis, we shall refrain from celebrating Weierstraß' successes; instead, we shall concentrate on his contributions to the basic concepts of analysis. However, regarding the concept of number, both happened to coincide.

© The Author(s), under exclusive license to Springer Nature Switzerland AG 2022
D. D. Spalt, *A Brief History of Analysis*,
https://doi.org/10.1007/978-3-031-00650-0_12

The Core of His Fame

It stands to reason that we shall not withhold a hint on the facts which constitute Weierstraß' renown as the "founder of rigorous analysis". It is based on Weierstraß' success in smashing the image of analysis of being—besides geometry— the second mathematical field which draws, by using clear concepts, convincing proofs from simple images. It was Weierstraß who (besides Bolzano, but this was unknown in the nineteenth century) designed analysis in such a way that it became a conceptual minefield, where each combination of notions produces surprises which render further investigations ever more difficult.

Earlier, analysts had already created grotesque structures. However, they were fenced in as extraordinary phenomena which were of no importance or could be controlled. In contrast to that, Weierstraß reversed things. He made these oddities the touchstone for the quality and the worth of the concepts instead of marginalizing them as meaningless anomalies.

Dirichlet in 1829: Analysis Has Frontiers!

In 1829, Dirichlet published a crazy idea: a "function" $f(x)$ which for each rational value of x has the "value" c and for each irrational value of x the "value" d, which is different from c.

There was *not one* concept of "function", including Dirichlet's own (p. 153), which was suitable for this creation. Although it has only two different "values" (in Dirichlet's thinking: equivalent to "ordinates"), namely c and d, this monster cannot be drawn. It cannot be drawn, because the rational numbers and the irrational numbers are intertwined with each other in such a way that they cannot be kept apart in a *drawing,* but only *as concepts.*

Each irrational number can be written as a decimal number with *infinitely many* digits which do not recur. If you stop this decimal number, say X, at *any* digit, you get a rational number X'. However, the *place* of the digit may be pushed as far out as one likes. The *difference* between X and X' is of magnitude 10^{-k}, if the number is cut off at the k-th decimal place. In other words: the difference between X and X' can be made as small as one wishes. In each case we have:

An irrational number has no nearest rational number.—The reverse is also true.

In the case of Dirichlet's so-called "function" this means: *near* each "value of the function" c there is a "value of the function" d—without the possibility of detecting the length of this "near". Consequently, it is *futile* to ask for the "jumps" of this "function" between its values c and d. Indeed, there is a distance between c and d (as they are assumed to be different), but the *smallest* difference of such values X and $\bar{\text{X}}$, which belong to c and d respectively, cannot be given. Or is it 0? So *no* difference?

It is true: Dirichlet published this so-called "function", but he did not really care about it. He only *stated* it. Why? Dirichlet wanted to say: there are cases in which you cannot "integrate"—analysis has limits. Without integral, no calculus. Dirichlet gave this "function" to demonstrate the *limits of the theory of analysis*. This strange object was not worth examining for him. And this is no wonder, because *actually* it is—according to his own concepts—no "function" at all.

Riemann 1854: These Limits of Analysis Can Be Shifted!

Dirichlet's student Bernhard Riemann was less humble. His idea was to fence in such extraordinary phenomena as proposed by his teacher. Riemann was not willing to delay the development of analysis. He tried to show that further progress was indeed possible and that it would not cause insurmountable problems.

In his habilitation thesis, written in 1854, Riemann constructed a function (and according to his concepts, see p. 155, it *actually is* a "function"!) which was not as unruly as that of his teacher, but which, nevertheless, was a hard nut to crack. Riemann presented a function which is "discontinuous" for incredibly many of its values, but nevertheless harmless—meaning that it can be integrated, and therefore qualifies as a legitimate object of calculus.

If Riemann's construction is reversed, starting from his integral, one gets an actual oddity, namely a derivative—a function!—which is "discontinuous" for *a great plenitude* of values, in Dirichlet's view: with *a great plenitude* of jumps. "A great plenitude" means: as many as there are fractions, and also ordered like them. The discontinuities of this derivative have the following properties:

1. They are infinitely many.
2. In between any two of them, there is another.

Such an enormous amount of discontinuities, that's a bit much! Nevertheless, this function can be integrated. With the help of a trick, Riemann guaranteed for the discontinuities: the closer they get, the more harmless they become. No doubt, this is a smart idea!

Weierstraß' Shocking Function

On 18 July 1872, Weierstraß presented a mathematical subject to the Royal Academy of Sciences in Berlin, which was suitable radically to transform the existing picture of analysis.

Two years after this event Leo Koenigsberger (1837–1921) still writes the following in his textbook with the title "Lectures on the Theory of Elliptic Functions":

> One of the main doctrines of analysis is the fundamental theorem that to each function of a real variable there really belongs a differential quotient, i.e. that the ratio of the increments of the function and of the variable can only be zero or infinite or is discontinuous because

of finite jumps in singular points of a finite length. However, for the rest it has finite values which are independent of the infinite small growth of the variable . . .

According to Koenigsberger, basically *each* function can be differentiated ("to it belongs a differential quotient")—apart from *possibly* a few singular values (at these "singular values" the derivative may not be defined).

This "main doctrine" or this "fundamental theorem of analysis" is *totally wrong*. This was shown by Weierstraß in his momentous lecture on 18 July 1872, where he proved the following:

> Theorem. *There exist continuous functions which cannot be differentiated at any single value.*

That Bolzano had already given such a function 40 years earlier (p. 113), remained unknown for another 40 years.

This theorem states the contrary of what Koenigsberger still says in his textbook 2 years later and which he calls one of the "doctrines" of calculus.

A. Koenigsberger *claimed* in 1874 that a continuous[1] function can *generally*—that means: for nearly all values—be differentiated.
B. Weierstraß proved in 1872: there exist continuous functions which cannot be differentiated *for any single* value.

A greater contrast is hardly imaginable.

The Aftermath of Weierstraß' Construction

Weierstraß had presented a shocking function; soon functions like that were called "capricious". In retrospect, we may say: analysis was at a crossroads. (i) Should one say: this we did not mean, such crazy "functions" we do not want, this counter-intuitive stuff we bar from analysis? (ii) Or should one say: okay, up to now we had a wrong idea of analysis—we must change our attitude and think *much more carefully* about the relations of our analytical concepts?

We remember that the exotic function which Dirichlet presented, rather casually in 1829, did *not cause any reaction*. Neither Dirichlet nor any of his contemporaries (as far as is known) responded in some way or another. Dirichlet's capricious function (as it was called later) was simply ignored in 1829.

43 years later the world had changed. *Not a single* leading analyst of that time thought of putting Weierstraß' oddity aside and to return to do business as usual. Quite the reverse, immediately the analysts of that time set out to create other exceptional functions. (Even a name was coined: "capricious".)

This started the search for a version of a "Value Analysis" which was as general as possible. Surely it is to Weierstraß' great merit that he triggered this development.

---

[1] We can assume that his omission of "continuous" was quite unintentional.

What is Weierstraß' Understanding of a "Function"?

In 1854, Riemann proposed at the very beginning of his doctoral thesis that a "function" is nothing else than this: to each "value" of an independent variable there corresponds *exactly one* "value" of the function (which is "the value of the function"—see p. 155). Today this opinion of "function" is generally accepted.

However, the victory of Riemann's concept of function was by no means guaranteed. It was Weierstraß who fought nearly all his life against this, according to his own opinion, wrong understanding of the subject. Weierstraß justified his objection in clear words. In his lecture on the foundations of the theory of analytic functions, in 1874, he stated:

> This general definition of the function is the same as the following geometrical one: a curve is a line which is not straight in any of its parts. From such a definition there cannot be deduced any positive property of the defined.

The last sentence explains what Weierstraß demanded from a *correct* definition of a mathematical concept: it must be possible to *develop* the relevant properties from the definition of an object. In short: *the definition has to designate the essence of that object.*

This coincided with the theory of definitions which Western philosophy had taught since Antiquity, i.e. those started by Aristotle or by Leibniz. (And, very interestingly, even by Georg Cantor, in 1883.) For clarity, I am going to coin a word for this basic view of mathematical thinking and will speak of a *"substancial"* definition in this case. Here, "substancial", one of the archaic forms of "substantial", draws attention to the fact that the given definition has theoretical substance.

Clearly, such a "substancial" definition does not drop from the sky. It has to be acquired through mathematical working. This was explained by Weierstraß quite explicitly in his lecture of 1874:

> It is impossible to define the concept of an analytical function in a few words—instead, this has to be developed little by little and this is the task of this lecture.

Obviously, the result of this Weierstraßian lecture cannot be given in a few words. Instead, we shall at least name the most important concepts which Weierstraß needed in order to define his concept of function. They are: (1) "polynomial", (2) "infinite series", (3) "uniform convergence" as well as "domain of convergence" and (4) "development of a power series centred at a value".—All of them being far from elementary concepts.

A Sudden Change

For nearly all of his scientific life, Weierstraß was thoroughly convinced that Riemann's concept of function was of no worth—precisely because it is *devoid of substance* and it is impossible to derive *any* important property of a "function" therefrom.

However, Weierstraß *recasted* his position in the very last lecture of his life on this topic, in 1886. At the age of 75 this mathematical Nestor changed his mind and advocated the opposite of his previous position. "Continuity" was the only property which he sustained, but besides this he subscribed to the supposition "that the function is a so-called *arbitrary:* it is lawless".

How come? A late revelation?

Not at all! Even later, in 1886, Weierstraß was still a proponent of substancial thinking. Instead, his change of mind regarding the *proper* definition of a "function" was changed by a basic mathematical result he had reached in the preceding year.

In 1885 Weierstraß had formulated and proved a profound theorem of analysis that still bears his name: the "Weierstraß Approximation Theorem".

> Theorem. *Each function which is continuous within a closed interval can be uniformly approximated by polynomials.*

In other words: Weierstraß had succeeded in drawing from the following, seemingly *simple,* two assumptions

(a) the domain is a closed interval, and
(b) the "function" (following Riemann) is continuous in this domain,

the *essence* of a "function". According to his view, the *essence* of a "function" is constituted by a power series, at least nearly. (This *nearly* is specified with the help of the technical term *"uniformly* approximated".)

Weierstraß was now prepared to accept Riemann's "empty" concept of function—for meanwhile he had succeeded in deducing from two seemingly *simple* additional phrases (finite and closed domain, continuity therein) an important mathematical content (uniform approximate representation of a function through a polynomial sequence). On Friday, 25th of June 1886, he announced the following definition:

> If to any system of values $(u_1, u_2, \ldots, u_n)$ there corresponds one and only one value x of the function, we call this a *unique[ly determined]* function of the variables $u_1, u_2, \ldots, u_n$.

To sum up: although Weierstraß revised his *mathematical* view, he retained his basic *philosophical* conviction. He continued to insist on *substancial* foundational concepts.

Weierstraß' Concept of Number

I(

We do not know any publication by Weierstraß on his concept of number. Nevertheless, his thinking about this concept is documented in some detail. This is due to two facts: first, he always stated his ideas on the concept of number at the beginning of his lecture cycle called "Preliminaries to the Theory of Analytic Functions" and secondly, there do exist several and sometimes very detailed notes taken from these lectures by students.

Nevertheless, it seemed impossible to figure out a clear-cut concept of (real and complex) numbers from these notes. That's why I proposed, in 1990, the view that Weierstraß worked on this concept all his life.

However, I have changed my mind since. Weierstraß does have a clear-cut concept of (real and complex) numbers. The real problem was: his students were unable to understand him and therefore they could not correctly report on it!

This is proved, for instance, by the lecture notes taken by Emil Strauß (1859–92) in winter 1880/81. They were discovered only recently and that by great chance: at the mathematical library at Frankfurt/Main university, in August 2016.

The Fundamental Hindrance That Obstructed the Understanding of Weierstraß' Concept of Number

Only after having completed the German original of this book in winter 2016/17, was I able finally to solve the riddle. The clue is this: Weierstraß relied on a concept not one historian of mathematics dared to think of: the concept of a (pure) set.

As we have noticed by now, Weierstraß was actually a conservative thinker, and set-theory was invented by his much younger pupil Georg Cantor (1845–1918) and by Richard Dedekind (1831–1916), from the second half of the 1870s onwards. Weierstraß never related to this theory.

Nevertheless, as Strauß' lecture notes clearly document, Weierstraß relied on the *concept* of set!

However, only on the *concept*—that means: Weierstraß did *not* make "set" the topic of mathematical reasoning. He did not invent set-theory. Weierstraß only *used* the concept of "set". Again and again he *proved* the associative and the commutative law for unions, as well as for products of sets *in special cases*—and that with the help of *analytical* arguments!

Nevertheless, Weierstraß' construction of the real as well as of the complex numbers is a true set-theoretical one. Which is to say: only *after* the invention of set-theory (in the twentieth century) Weierstraß' construction of real numbers *developed* into being *the most elementary we know today.*

Yet Weierstraß' students *had no chance* of understanding him. For, not surprisingly, Weierstraß never placed any emphasis on the concept of set. He did not even coin a name for it. Why should he? In his thinking it was a mere auxiliary concept and, indeed, a very *trivial* one. Why emphasize it? (Again, an excellent mathematician is not necessarily an excellent teacher.)

Consequently, this new concept *escaped* the understanding of his students—and what is not understood will not be adequately recorded. Normally.

The Peculiarity of the Student Emil Strauß

The student Emil Strauß was somewhat peculiar. In his lecture notes he did not (only) write down what he really understood (and valued as important)—but plainly *everything* he thought he might grasp. In one, very curious, case his writing actually amounts to *complete mathematical vacuity*—nevertheless this passage is highly valuable for the historian because it shows a sharp attack by Weierstraß on his colleagues; whereas parallel lecture notes, taken in the same lecture by other students (among them e.g. Adolf Kneser (1862–1930)) do not even show a hint of this attack.

Thus Emil Strauß' notes contain some sentences which clearly *prove* that Weierstraß operated with (pure) sets. You only have to take these single sentences *literally*—which is *not* an easy task! It took me a considerable amount of time to understand this. However, if the result finally adds up to a coherent mathematical statement, one can be fairly certain that one is embracing Weierstraß' idea.

Weierstraß' idea is truly simple: he takes the utmost possible generalization of decimal numbers!

Preliminary

This generalization is carried out in three easy steps, which all follow this assumption:[2]

> Preliminary: *We take the natural numbers as well as the fractions as given.*

This Preliminary comprises Weierstraß' utmost precise construction of the natural numbers as well as the fractions from scratch. This foundational part is a philosophical construction, and as our topic is analysis—not foundations—, we have to omit this here. However, I cannot resist the temptation of disclosing my judgement: Weierstraß' *philosophical* construction of the natural numbers is, in my opinion, the most precise which is known today, including that of Edmund Husserl (1859–1938), who attended Weierstraß' lecture in winter 1880/81.

The Construction

We proceed with the construction of Weierstraß' real numbers:

---

[2] As already announced as well as justified in the introduction (see p. xxii), in the following I do not stick to my historiographical method, but instead I present a former idea completely in modern language.

> 1. Disregard the sign and start with the interpretation of a decimal number $d_0.d_1d_2d_3\ldots$ to be a (sometimes infinite) *set* of fractions: $\left\{\frac{d_0}{10^0},\ \frac{d_1}{10^1},\ \frac{d_2}{10^2},\ \ldots\right\}$
>   Now we go beyond this and allow that
> 2. the nominators and
> 3. the denominators can be *any* natural number.

That's all! These three steps give us the utmost general concept of "irrational" number, *together* with the two *direct* operations "addition" und "multiplication". We now have the following

> Definitions (*addition, multiplication*).
> 1. The "addition" of two such numbers is their set-theoretical union— followed, of course, by the addition of each two elements with the same denominator.
> 2. The "multiplication" of two such numbers is, what we may call the "Weierstraß Product": multiply each element of one number with each element of the other number and finally add in the resulting set all fractions with equal denominator.

These two operations are well-defined. They are associative as well as commutative and they even obey the distributive law—because these laws hold for sets and the natural numbers and their operations.

This was truly easy, wasn't it? Unfortunately, we are not yet through!

What is missing? Well, we have the set of *all* imaginable (real) numbers (without sign) you may dream of. We are able to add and to multiply them—but do not know how to *compare* them yet! But of course, we *need* equality as well as comparisons like, e.g.:

$$\left\{\tfrac{1}{3}\right\} = \left\{\tfrac{3}{10^1},\ \tfrac{3}{10^2},\ \tfrac{3}{10^3},\ \ldots\right\} < \left\{\tfrac{1}{2}\right\} \ !$$

Further Preliminaries for Tackling a Real Difficulty

How shall we deal with this plenitude of numbers? Where should we begin?

Since Weierstraß' numbers are finite as well as infinite sets, we shall start with the finite ones. Weierstraß himself gave *no one name* to his numbers, but we shall do. As the reader may expect:

> Definition (*fraction*). Each number which is a finite set, will be called a "(general) fraction".

If we were pedantic we would choose the name "general fraction", for we already *have* constructed the "fractions" in our initial *Preliminary*. However, things seem fairly clear and therefore we might be allowed to dispense with this "general". Thus, let us stick for one more moment with the attribute "general", for we have also to state the following:

> Definitions (*equality, transformation*).
>
> 1. Two (general) fractions are "equal", if one of them can be *transformed* into the other.
> 2. A "transformation" of a (general) fraction is any finite series of changes of one or more of its elements (a) either by cancelling or its opposite, or (b) by splitting it into two or more fractions or *vice versa*.

Therefore, we have e.g. $\left\{\frac{1}{2}, \frac{1}{4}\right\} = \left\{\frac{5}{7}, \frac{1}{28}\right\}$, since

$$\frac{1}{2} + \frac{1}{4} = \frac{3}{4} = \frac{21}{28} = \frac{5}{7} + \frac{1}{28}$$

from the *Preliminary*. Similarly, we state:

> Definition (*less than*). A (general) fraction a is "less than" ($<$) another number b if b can be transformed such that $a \subset b$.

Caution: you need to transform b, for we need to have

$$\frac{1}{3} < \frac{1}{2} = \frac{2+1}{6} = \frac{1}{3} + \frac{1}{6},$$

and this you cannot get by transforming a. Please, remember: this definition is not an arithmetical, but a set-theoretical one!—It is true: Weierstraß went for the arithmetical version, not for our set-theoretical one.

We opted in favour of the set-theoretical definition for "less than", because it is Weierstraß' solution in the case of the infinite sets. The real difficulty of this number concept is to define equality and to make comparisons for infinite sets—because the condition to permit only finitely many transformations is here not enough. We need to have

$$\left\{\frac{1}{1}\right\} = \left\{\frac{1}{2}, \frac{1}{4}, \frac{1}{8}, \ldots\right\} \quad \text{as well as} \quad \left\{\frac{1}{6}, \frac{1}{12}, \frac{1}{24}, \ldots\right\} < \left\{\frac{1}{2}, \frac{1}{4}, \frac{1}{8}, \ldots\right\},$$

but one *cannot* carry out infinitely many transformations.

Solution of the (Perhaps) Real Difficulty

The concepts of equality and comparison in cases where numbers which are infinite sets are involved, seem to be the real difficulty of this number concept. These definitions do not come easily.

Really? Don't we have the fractions (the finite sets)? Doesn't it suggest that we should take them to be our standard?

To make his students aware of the difficult notion of equality in the case of infinite objects (sets), Weierstraß presented the following "fallacy" (as he called it): starting with

$$x = \tfrac{1}{2} + \tfrac{1}{4} + \tfrac{1}{8} + \dots,$$

one gets

$$x + 1 = 1 + \tfrac{1}{2} + \tfrac{1}{4} + \tfrac{1}{8} + \dots = 2x,$$

and this implies

$$x = 1.$$

Weierstraß comments: "This conclusion uses a 'law of equality', which has not yet been stated." (Interestingly, only the lecture notes of Emil Strauß document this passage—and they show the (obvious) mistake "$1 + x$" instead of "$2x$"—, whereas e.g. the notes of the listener Adolf Kneser do not.)

Be that as it may, Weierstraß already had an idea how to solve this problem, and according to the sources, he had it already in, or (presumably) even before, 1868. That was at least 4 years earlier than Dedekind—who published the same idea, albeit with quite a different interpretation. The idea is this:

Definitions (*equal, less than*). A (general) fraction f is "less than" a number q, if there exists a finite subset $g \subset q$ with $f < g$. Two numbers are "equal", if each fraction less than one is also less than the other; if only one part of this condition is fulfilled, the relation "less than" ($<$) *resp.* "greater than" ($>$) is defined.

In 1872, this was Dedekind's idea of *defining* the (irrational) numbers (see p. 202). However, Weierstraß had used this idea to *compare* the (irrational) numbers already in 1868 or even earlier. We see the difference in their ways of thought in those approaches: Weierstraß used the idea of *comparing* his numbers, which he had previously defined (Weierstraß is a "substancial" thinker!), while Dedekind used this relation to *define* his numbers (therefore, I propose to call him a "relational" thinker).

The arithmetical laws for these relations ($=$, $<$) are easily proved.

Weierstraß abhorred infinite numbers (he sided with Leibniz and Cauchy against Johann Bernoulli and Euler) and proposed the

Definition (*finite*). A number a is called "finite," if there exists a natural number n $\left(:= \left\{ \tfrac{n}{1} \right\} \right)$ with $a < n$.

(Weierstraß did not even dream of *infinite* natural numbers—at most in nightmares. However, what would Johann Bernoulli have said, see p. 52?)

And then Weierstraß banished the infinite numbers by calling them all together "∞".—Consequently, ∞ is not a true number, as it violates the arithmetical laws.

The Benefit of Structural Thinking

If we add the number $\left\{\frac{0}{1}\right\}$, these finite Weierstraßian numbers constitute what we call nowadays a *monoid*. Actually, we even have two monoids: one for each of the two operations, addition and multiplication.

Today, at least in our second semester at university, we are taught how to enlarge a monoid $(M, \circ)$ to a group $(G, \bullet, \star)$: we just take G to be $M \times M$ and operate this way:

$$(a, b) \bullet (c, d) := (a \circ c, b \circ d),$$

$$(a, b) \star (c, d) := (a \circ d, c \circ b),$$

$$(a, b) = (c, d) :\Longleftrightarrow a \circ d = c \circ b,$$

where "$\star$" is the "inverse" operation of "$\bullet$".

> Actually, this is just what Weierstraß did with his finite numbers, taking them to make up our set M! Exactly in this way he constructed his "real" numbers to be pairs of his "irrational" numbers. And since his "irrational" numbers M constitute two monoids, he actually got a *ring $G = M \times M$ with the unit 1.*

Heavy stuff, indeed! However, for those familiar with elementary modern algebra (*after* Bourbaki), It's mere child's play. Nonetheless, it's mere really astonishing that Weierstraß managed to come up with this construction, without *any* algebra to hand!

Clearly, not one of his students, not even the most talented, was able to grasp this construction. Besides, the concept of "ordered pair" had not yet been created (this only happened in the twentieth century). The same holds for the use of *arbitrary symbols* for operations. Inevitably, Weierstraß' students were unable to distinguish between our "$\circ$" and "$\bullet$" from above, and, even more dramatically, they confused the formal "$\star$" with the arithmetical ("true") "$-$"—because Weierstraß always used "$+$" as well as "$-$"! Weierstraß was not a protagonist of mathematical revolutions, but he was an enormously rigorous thinker.[3]

That is why we could not access this idea before Emil Strauß' lecture notes where found (in August 2016) and my subsequent interpretation. It was very lucky that Weierstraß disclosed his view of the numbers *in* (nearly) *all its elementary details* in winter 1880/81—and this in the presence of this ambitious student. Of course, Weierstraß did not go into all these details in each of his foundational lectures!

---

[3] In case he knew that $a = \bar{a} + c$ as well as $b = \bar{b} + c$, he easily concluded $(a, b) = (\bar{a}, \bar{b})$ for all irrational numbers a, b. The decisive point is, that he already *knew* the first two equations—for otherwise he would have used *general* subtraction of his irrational numbers, which is impossible.

The Benefit and the Disadvantage of Structural Thinking

Well, even a Weierstraß was not perfect and could make an error. Weierstraß taught procedures to divide his irrational as well as his real numbers—in fact, in different lectures he proposed two different methods.

However, our structural analysis of his construction genuinely proves: they are both wrong. The reason is that this idea of construction only produces a ring, not a field. And it is true: both of the methods lectured by Weierstraß are erroneous. They both rely on the possibility of *subtracting* "irrational" numbers—but this is *impossible.* (In hindsight: *if* it were possible, it would not have been necessary to construct the operation of subtraction in a *pure formal way*—as it is done by taking $M \times M$ which Weierstraß actually did.)

This line of thought inevitably leads to the question: *Do Weierstraß' real numbers form a ring (with* 1 *as unity) or a field?* We may put it this way: *can Weierstraß' real numbers be divided, yes or no?*

This question is very productive. It can be answered in two ways:

1. The real numbers are objects—it does not matter how they are constructed. These objects can be taken as subjects of a structural analysis. (You may construct the integers by adding the "signs" $+$ and $-$ to the natural numbers and zero—nevertheless, we obtain the structure of a group from this monoid.)
2. Whether some objects form a certain structure *depends* on the definition of "operation" taken.

This is a book neither on algebra nor on philosophy of mathematics, but solely one on the history of analysis. Therefore, we have to limit ourselves to elementary aspects of this topic and merely state two things:

- There is a fundamental difference between direct operations and their reverse, the inverse operations. The latter are methods of trial and error, they *cannot* be carried out directly.

  This fundamental difference between these two kinds of operation is smashed by taking the algebraic point of view. It makes a great difference whether we say: "An *operation* is a method of generating a new object from two given ones." or if we state: "An *operation* is defined whenever there exists an object which fulfils certain conditions." The first standpoint is Weierstraß', the second is Hilbert's.

- We all know how to subtract and how to divide decimal numbers (we just take suitable approximations). We have learnt this in school. Consequently, we are all able to subtract and to divide Weierstraß' real numbers—we only need to take them in their decimal (or binary or some equivalent) form.

  We may put this in higher algebraic language and say elaborately: (i) Equality is an equivalence relation for Weierstraßian real numbers. (ii) In each equivalence class there exists a decimal (or ...) number—which in Weierstraß' sense may be called the "value". (iii) The decimal (or ...) numbers are complete. (This proof is due to Hilbert.)

Therefore, we answer the above question in a personified way:

> In Weierstraß' view, his real numbers do not constitute a field, whereas according to Hilbert's view, they do.

(That Weierstraß himself, erroneously, held his real numbers to be a field, is a historical fact.)

Postscript

It was as late as 1994 when John Horton Conway (1937–2020) characterized the easiest way of constructing the real numbers:

> Proceed from the natural numbers to the non-negative rationals (or the strictly positive ones if you prefer), then construct the non-negative (or positive) reals from these, so having no sign-problem, and then construct signed reals from these in the way that we constructed the signed integers from the natural numbers. I think that this is in fact the simplest way to construct the real numbers along traditional lines.

Nevertheless, neither Conway nor any other mathematician ever since came up with this construction of the real numbers—although, as we have seen, it is *really* a straightforward generalization of the decimal numbers.

The inevitable conclusion is, yet again, the well-known truism: *Mathematics is* REALLY *difficult!*

l)

Infinite Series

With his construction of the *general* concept of "quantity"—which is our "real number"—Weierstraß was able to do analysis. In particular, he built the concept of "infinite series" or simply "series".

Weierstraß repeated the method used for his numbers step-by-step. He started with the study of "infinite series" of "fractions", followed by those of "irrational numbers". That is, he considered the series

$$a_1 + a_2 + a_3 + \dots ,$$

where a_k are irrational numbers. Each of it has the form

$$a_k = \alpha_0^{(k)} \cdot 1 + \alpha_1^{(k)} \cdot \frac{1}{n_1} + \alpha_2^{(k)} \cdot \frac{1}{n_2} + \dots$$

wherein the order does not matter. Taking σ_i to be the sum of all multiples $\alpha_i^{(k)}$ of the fractions $\frac{1}{n_i}$ for all the k terms of the series (i.e. $\sigma_i = \bigcup_k \alpha_i^{(k)}$) one gets

$$s = \bigcup_k a_k = \bigcup_i \sigma_i \cdot \frac{1}{n_i} \, .$$

(Of course, Weierstraß did not use the symbols of set-theory—this is our modern reading of his original concepts.) Therefore, to assure the finiteness of the series Weierstraß demands:

1. *All irrational numbers a_k of the series must be finite.* (Clearly, this demands that in all summands $a_k = \bigcup_i \alpha_i^{(k)} \cdot \frac{1}{n_i}$ the *numerators* $\alpha_i^{(k)}$ of the fractions with denominator n_i have to be finite; however, this condition is only necessary but not sufficient.)

2. Each "element" of s (which is to say, each $\sigma_i \cdot \frac{1}{n_i}$) has a finite multiplicity σ_i, or: *all σ_i are finite.*

Subsequently, this "sum" $s = \bigcup_k a_k$ is made up as follows: in each "element" the numerator of the fraction with denominator n_i has to be σ_i, which was built from all the irrational numbers a_k of the series.

The well-known example $1 \cdot 1 + 1 \cdot \frac{1}{2} + 1 \cdot \frac{1}{3} + \dots$ shows, that the given conditions are not sufficient. Therefore, the relevant question is:

In which cases is the series $s = \bigcup_k a_k$ of the irrational numbers a_k finite?

To answer this question, we need to know Weierstraß' concept of the "summability" of a series.

Summability

Since Cauchy we are used to defining the concept of "convergence" of an infinite series

$$u_0 + u_1 + u_2 + \dots = s_n + r_n$$

to be the condition

$$\lim r_n = 0 \, .$$

Weierstraß gave another concept, and, to make things more clear, he also chose another name: "summability" ("Summierbarkeit"). His notion is this:

Definition (*summability*). An infinite series

$$a_1 + a_2 + a_3 + \dots$$

is called "summable" if its sum is finite (*see* p. 177).

As we already know: the sum s of the series $\bigcup_k a_k$ is $\bigcup_i \sigma_i \cdot \frac{1}{n_i}$.

Unfortunately, Weierstraß did not state this definition. We have to deduce it from the text or, more precisely: from the name he used for the concept.

Summability for Series of Irrational Numbers

Weierstraß gives a *criterion* for the sum of an infinite series to be finite. In the case of a series of irrational numbers, he says (his style is moderately updated):

> Theorem. *The infinite series $\bigcup_k a_k$ of irrational numbers a_k is finite if and only if it is possible to determine a number g, which is larger than each sum of finitely many a_k.*

Weierstraß proves this theorem by demonstrating it to be "correct". That means: if and only if the stated criterion is fulfilled, then $s = \bigcup_k a_k$ is finite.

Clearly, if the sum of the infinite series is finite, the given criterion is valid. But what about the opposite direction?

If the specified criterion is fulfilled, we shall get the sum s of the series to be $s = \underline{\lim} g$, i.e. the least upper bound of $\bigcup_i a_i$. However, Weierstraß did not say a single word about how this sum can be defined. That is why his proof of the above theorem cannot be understood that easily. It plainly lacks the following definition:

Definition (*sum of an infinite series, summability*). The *sum* of an infinite series is a (finite) number which is the least upper bound of any finite sum of its terms.

Nowadays we are to take this definition to be the specification of Weierstraß' definition of "summability".

Weierstraß even proves the following:

Theorem. *If the infinitely many irrational numbers a_k are arbitrarily grouped, and the sums of these groups are built, the result will always be the same.*

In the course of the long proof, some important theorems on irrational numbers are revealed.

Theorems on Irrational Numbers

Theorem. *Each finite irrational number a, with infinitely many elements $\alpha_i \cdot \frac{1}{n_i}$ can be represented in this way:*

$$a = a_0 + a_1 \,,$$

where (i) a_0 has finitely many elements; (ii) a_1 can be made arbitrarily small.

This statement may be expressed this way: each finite irrational number a can be approximated by a fraction (a_0) to any degree of accuracy (a_1).

This leads Weierstraß to his theorem for the "equality" of two finite irrational numbers:

> Theorem. *Two finite irrational numbers a and b are equal if and only if it is possible to split each of them into two: $a = a_0 + a_1$ and $b = b_0 + b_1$ (a_0 and b_0 being fractions) such that $|a_0 - b_0| < \varepsilon$ as well as $a_1 < \varepsilon$ and $b_1 < \varepsilon$ are valid for any previously given ε.*

(It is easy to see that the subtraction of a (general) fraction is possible.) Stated without formulae:

> *Two finite irrational numbers are equal if and only if it is possible to split each of the quantities into two, an approximating fraction and an (irrational) excess, such that the difference of these approximating fractions as well as each of the two remainders in itself is less than any previously given tolerance ε.*

Formulated this way, the theorem seems to be clear. The technical proof in the lecture notes comprises one large page. Relying on this theorem, Weierstraß "easily" proves the following

> Theorem. *In an infinite sum, equal irrational numbers can replace each other.*

Summability for Series of Real Numbers

The addition of real numbers is different from that of irrational numbers. (Conway did not ponder on that!) Therefore, in case of series of real numbers another *criterion* for summability is required. Weierstraß presents it as follows:

> Main Theorem. *If the series x_1, x_2, x_3, ... has the property that there exists a finite number g such that the absolute sum of arbitrarily many and arbitrarily chosen terms is less than g, then the series is summable.—The reverse also holds.*

To prove this, Weierstraß argues thus:

(a) Take the positive of the chosen terms x_k (i.e. the pairs $x_k = (a_k, b_k)$ of irrational numbers a_k, b_k with $a_k > b_k$). One can change each of them in such a manner ($(a_k, b_k) = (\overline{a_k}, \overline{b_k})$) that its "negative" part $(\overline{b_k})$ is as small as one wishes.

(b) The same can be done with the negative terms x_k (i.e. the pairs $x_k = (c_k, d_k)$ with $c_k < d_k$) such that $x_k = (\overline{c_k}, \overline{d_k})$ where $\overline{c_k}$ is as small as one wishes.

(c) Consequently, it is possible to change the positive of the chosen terms in such a manner that their "negative" parts $(\overline{b_k})$ are less than the terms f_k of a converging series $\sum f_k = f$.

(d) The same can be done with the negative of the chosen terms: their "positive" parts $(\overline{c_k})$ can be made less than the terms h_k of a converging series $\sum h_k = h$.

(e) Consequently, the series can be replaced by (i) a sum of positive $(\overline{a_k})$ (ii) as well as "negative" fractions $(\overline{d_k})$ plus (iii) a sum of the "negative" irrational numbers $(\overline{b_k})$, which if taken absolutely $(\overline{b_k})$, have a finite value $(< f)$, as well as (iv) a sum of irrational numbers $(\overline{c_k})$, which is also finite $(< h)$. (This is true even if the series is not summable!)

(f) However, the criterion demands that there exists a finite number g such that the absolute sum of any finite set of terms x_k is less than g. Consequently, each of these sums is less than $(g + f) + (g + h)$.

   (g) Therefore, each absolute sum (of finitely many) terms of the series is guaranteed to be
      less than $2g + f + h$.
   (h) From this, Weierstraß concludes that the series is summable—which proves the Main
      Theorem.

As Weierstraß operates with the absolute value of the partial sums, his last
conclusion relies again on the (unsaid) concept of the least upper bound, as it has
been in the case of series of irrational numbers.

The Concept of "Convergence"

Weierstraß uses the notion of "convergence" only towards the end of his lecture
(on p. 155 of Strauß' notes). There he even *identifies* it with his concept of
"summability".

When speaking of "convergence", Weierstraß relies of course on the *order* of
the terms in the series. After having introduced the "complex" numbers, Weierstraß
proves the following:

> Theorem. *If the (complex) series is summable and has the sum s, it is possible to determine
> to any given ε* [4] *a natural number n such that*

$$|s - s_n| < \varepsilon .$$

And he adds: "Because of this theorem one calls a summable series also convergent
and says that the sum s_n converges to the value s."

Weierstraß continues: if the concept of "limit" of a series is defined, this notion is
taken to be equivalent to the concept of its "sum". However, the value of this "sum"
may depend on the order of the terms. Series with this property Weierstraß suspends
"for the time being", closing with the following

> Definition ((*un*)*conditional convergence*). If the order of the terms influences the value of
> the sum, the series is called "conditionally convergent"; otherwise it is called "uncondi-
> tionally convergent" (or "absolutely convergent"—in case of real numbers both notions are
> equivalent).

Upshot of Weierstraß' Concept of Number

Weierstraß takes the concept of "series" to be a generalization of the concept of
"sum". That is why he dispenses with the idea of order when dealing with series.

Therefore, his notion of the finiteness of the sum of a series, namely "summability", is narrower then our concept of "convergence" (which demands *ordered*
series). Consequently, Weierstraß' "summability" coincides with our concept of
"unconditional convergence".

---

[4] Weierstraß, indeed, uses "δ" ...!

Results:

1. Weierstraß succeeds in founding analysis on a well-established concept of (irrational, real and complex) number.

2. Additionally, Weierstraß succeeds in formulating a substancial concept of "equality" for "irrational" numbers.

3. In his untimely anticipation of an algebraic construction Weierstraß completes the monoid of his irrational numbers to a group, thereby giving a unique elementary construction of the real numbers, which has not been repeated since.

4. Because Weierstraß took the concept of "series" to be a generalization of the concept of "sum", his notion of finiteness for the sum of a series (called "summability") is narrower than the usual concept of "convergence" used nowadays: "summability" is "unconditional" convergence.

Weierstraß in Retrospect

Weierstraß was not the first to define clear concepts of analysis, nor did he invent epsilontics. He learnt all of that.

Nevertheless, he pioneered other foundational landmarks in analysis:

1. In his "Approximation Theorem" Weierstraß succeeded in developing calculatory expressions ("polynoms") as a rather good approximation for any "continuous" function, which has a "compact" domain. This paved the way for Riemann's concept of a function in analysis.

2. Weierstraß actually developed his concept of the real numbers *constructively*. Albeit, this simple and elegant construction came far too early and therefore could not be expressed in a suitable language. This is also why his students, even the most gifted, were unable to grasp it. Mysteriously, this concept was not re-invented in fourteen decades.

3. Weierstraß' idea of turning "capricious" functions into legitimate objects of analysis paved the way for "Value Analysis" finally to become a very general theory of "Functional Analysis".

Weierstraß' thinking was clearly traditional—I proposed the name "substancial". He was not keen on manipulating formulae: quite contrarily, he substantiated every step within his construction of the numbers.

And it is a real paradox of history that Weierstraß was (to my knowledge) the very first mathematician to give a strictly formal construction of a group by forming the Cartesian product of a monoid with itself—even decades before the concept of "ordered pair" was created, not to speak of the concept of a "monoid".

Today's judgement is inevitably this: Weierstraß has given a precise concept of real number; (i) it is totally systematic, (ii) it does not rely on the concept of "negative" number and (iii) it is the most elementary known to this day.

However, nothing originates from nothing. Weierstraß has to pay a price for his concepts to be of such an elementary character. This price is the lack of inverse operations: Weierstraß' irrational numbers *cannot* be subtracted nor divided (see p. 179). To implement the inverse operations for Weierstraß' irrational numbers one has to change from his most elementary concept to a more ambitious one, to a "positional system" of numbers. But the way of subtracting and dividing decimal numbers we are all taught in school.

Literature

Cantor, G. (1883). Über unendliche lineare Punktmannigfaltigkeiten (No. 5: Grundlagen einer allgemeinen Mannigfaltigkeitslehre). *Mathematische Annalen, 21*, 545–586. cited following Zermelo 1932, pp. 165–209.

Conway, J. H. (1994). The surreals and reals. In Ehrlich, P. (Ed.). *Real numbers, generalizations of the reals, and theorems of continua*. Synthese library, Studies in epistemology, logic, methodology, and philosophy of science (Vol. 242, pp. 93–103). Dordrecht, Boston, London: Kluwer Academic Publishers.

Dirichlet, G. L. (1829). Sur la convergence des séries trigonométriques qui servent à représenter une fonction arbitraire entre des limites données. *Journal für die reine und angewandte Mathematik, 4*, 157–169. See also Kronecker & Fuchs (Eds.) 1889, 1897, vol. 1, pp. 117–132.

Hettner, G. (1874). *Einleitung in die Theorie der analytischen Funktionen von Prof. Dr. Weierstrass, nach den Vorlesungen im S[ommer]s[emster] 1874.* mimeographed by the Mathematical Institute of the Göttingen university (1988).

Kneser, A. 1880/81. *Weierstrass, Einleitung in die Theorie der analytischen Funktionen.* Staats- und Universitätsbibliothek Göttingen, manuscript, signature Cod Ms A Kneser B 3.

Koenigsberger, L. (1874). *Vorlesungen über die Theorie der elliptischen Functionen.* Leipzig: Teubner.

Kronecker, L. (vol. 1) & Fuchs, L. (vol. 2) (Eds.) (1889, 1897). *G. Lejeune Dirichlet's Werke.* Berlin: Georg Reimer.

N. N. 1886. *Ausgewählte Kapitel aus der Funktionentheorie,* Vorlesung von Prof. Dr. Karl Weierstraß, Sommersemester 1886. In: Reinhard Siegmund-Schultze (Ed.), vol. 9 of the series *Teubner-Archiv zur Mathematik.* Leipzig: Teubner, 1988.

Strauß, E. (1880/81). *Weierstrass, Einleitung in die Theorie der Analytischen Functionen.* University of Frankfurt am Main, archive, file number 2.11.01; 170348.

Zermelo, E. (Ed.) (1992). *Georg Cantor. Gesammelte Abhandlungen mathematischen und philosophischen Inhalts.* Hildesheim: Georg Olms Verlagsbuchhandlung, reprint 1966.

Chapter 13
Analysis' Detachment from Reality—and the Introduction of the Actual Infinite into the Foundations of Mathematics

A Pessimistic Mood

The Initial Situation in 1817

Bernard Bolzano was the first to introduce the concept of value into analysis in 1817 (pp. 101f). "Values" are, first of all, numbers as well as (usually) $\pm\infty$ ($\pm$ infinity). However, what *is* a number, *precisely?*

Basically, numbers arise from *counting:* 1, 2, 3, ... Anyhow, that is not enough for analysis. From the beginning, analysis was there to *measure:* areas, volumes, gradients, curvatures, etc. But exact measurement needs more than just the *natural numbers* 1, 2, 3, ...; additionally, it requires at least the *fractions:* $\frac{1}{2}, \frac{1}{3}, \frac{1}{4}, \ldots$

However, the surveyors of classical Greece already knew:

> Even fractions do not suffice for exact mathematics.

For example, it is impossible to express the exact length of the diagonal of a square with side 1 in fractions. (This proof is very instructive and not complicated, but it does not belong to analysis and so it is omitted here.)

In 1821, Cauchy repeated the age-old saying of the mathematicians:

> "Number" is the ratio of two quantities.

This only transfers the problem to the question: what are "quantities"? but does not solve it.

© The Author(s), under exclusive license to Springer Nature Switzerland AG 2022
D. D. Spalt, *A Brief History of Analysis*,
https://doi.org/10.1007/978-3-031-00650-0_13

Failure

For a long time, there was no progress regarding this issue. Since 1817, Value Analysis progressed rapidly in many fields—but not in that of the concept of number. No new idea stirred up the minds of the leading mathematicians. (Bolzano tried a proposal; however, judged as a radical political thinker, he had been dismissed from the academic world, and, consequently, his ideas, which were hidden in unpublished manuscripts, remained unknown in the nineteenth century.)

Obviously, it *was* a problem. In the foregoing chapter, we have seen that Weierstraß worked in order to solve it. Unfortunately, his solution was not understood and consequently remained unknown.

In 1867, half a century after Bolzano's important treatise, the gifted student of Riemann, Hermann Hankel (1839–73), stated in a book the following pessimistic judgement:

> Every attempt to treat the irrational numbers in a formal manner and excluding the notion of [extensive] quantity itself is bound to lead to the most abstruse and laborious artificialities which—even if they could be carried out with complete rigour and we have every good reason to doubt this will be possible—will not be of any favourable scientific value.

(The word "extensive" is added to elucidate Hankel's intention.)

In other words: in 1867 Hankel was convinced that for any "foundation" of a suitable analytical number concept, the recourse to intuition is inevitable—even if it *were* to be possible to give a *pure* notion of number (that is one without any help of intuition), this might lead to such complications ("abstruse", "artificialities") that it would be of no help to science.

1872: Two at Once

Only 5 years later, Hankel's conviction had been overcome and proven to be wrong. Within an interval of only a few weeks, two (*completely different*) constructions of the real numbers were published. Both constructions avoided any recurse to intuition but instead were purely conceptual. The most elementary one, Weierstraß' idea (see pp. 174f), remained unknown until today. That is why these two constructions are the ones still taught to students today. Usually just one of them, because *one* definition of a concept is quite enough. Besides, on closer examination, it turns out that both concepts of number amount to the same.

One generation later, in 1899, David Hilbert found a way of characterizing this concept of number. Hilbert set up a *system of axioms* that lists all *essential* properties of the real numbers. Now logicians were able to prove that any two systems of numbers that fulfil those axioms are interchangeable. In other words: *strictly speaking*, there exists only one single system of numbers that has *exactly* these properties. We already hit upon this (pp. 179f), and we will return to it yet again (pp. 205f).

The inventors of these constructions of the "real" numbers are Georg Cantor (who went on to invent set-theory) as well as the triumvirate Traugott Müller (in 1838), Joseph Bertrand (in 1849), and Richard Dedekind (in 1872, he also invented set-theory). In the following, we will outline these two constructions.

Georg Cantor's Construction of the Real Numbers

The Situation

First, we have to mention briefly Heinrich Eduard Heine (1821–81). The older Heine was working together with the younger Georg Cantor at Halle University, and as a consequence, each of them published a treatise on this topic at the same time. Cantor's paper was actually on another subject, and thus it contained the great news only roughly. In contrast to this, Heine tackled the issue fundamentally and pedantically went into all the details.

Aiming at a comprehensive presentation, Heine outlined in his introduction the general situation of analysis in 1872. According to him, it was like this (all *emphases* are his):

> The progression of the theory of functions is essentially hindered by the fact that some elementary theorems, although they are proved by an astute scientist, are still doubted. From this it follows that research results, which are based on those theorems, are not generally held to be valid.
>
> A possible, and very likely, explanation is the fact that the principal teachings of Mr Weierstraß disseminate in wide sections, directly via his lectures and through oral informations as well as indirectly through copies from lecture notes which originate from his courses. But they have not yet been published by himself in an authentic version. Consequently, there exists no place where one can find the theorems *gradually and systematically developed*.
>
> However, the truth of these theorems relies on the not completely established definition of the irrational numbers, which is frequently confusedly and negatively affected by geometrical ideas, particularly by the *generation* of a line *through motion*.
>
> *These theorems are valid if the definition of irrational numbers given below is assumed to be their foundation: where numbers are called equal if they do not differ by any given number, being as small as one wishes, and where, in addition, the irrational number is due to actual existence,* such that a function has for each *single* value of the variable a *definite* value, too, be it rational or irrational.
>
> However, if one took another point of view, there could be very good reasons to make objections to the truth of the theorems.

This extraordinarily lucid passage aptly formulates (like Cauchy's preface to his textbook of 1821 at that time) the situation of analysis in 1872, as well as the *possible* changes caused by this new construction. "Possible", because Heine *explicitly* left open, whether the community of mathematicians will accept this new idea of his fellow mathematician Georg Cantor or not. Obviously, Heine did not take the mathematical/political success of this construction for granted—although he felt capable of demonstrating its mathematical force (and actually did so).

Cantor's Construction Arises Out of Weierstraß'

Georg Cantor (1845–1918) heard Weierstraß' lectures. However, Cantor was not Weierstraß and therefore he thought differently. This *other way of thinking* produced an idea, which even today, one and a half centuries later, belongs to the fundamentals of analysis. What is this idea?—A surprising one!

According to Weierstraß (we did not go into all the details), a real number is a pair of two irrational numbers (*cf* p. 178):

$$\begin{cases} k_0 * 1 + k_1 * \frac{1}{n_1} + k_2 * \frac{1}{n_2} + \dots \\ l_0 * 1 + l_1 * \frac{1}{m_1} + l_2 * \frac{1}{m_2} + \dots \end{cases}$$

What is written here as "1" or "$\frac{1}{n_i}$", "$\frac{1}{m_i}$" was thought by Weierstraß to be a "unity" *resp.* its "exact parts", whereas the "k_i" and "l_i" are their "multiplicities". Each of the above lines is a distinct part of the "real number", the first being called "positive", the second "negative"—yet with a very special meaning (see p. 183). And, last but not least: for Weierstraß, the + was a priori commutative, even in case of infinitely many elements: *Weierstraß thought an irrational number to be a* SET—a concept completely unknown to his contemporaries in the early 1870s.

Cantor introduced a *fundamental change* into this construction. In my opinion: *Cantor plainly misunderstood Weierstraß.*

1. He completely dismissed the concept of "unity" (as being a useless philosophical notion). In other words, Weierstraß' "unities" "$\frac{1}{n_i}$" were taken by Cantor to be mere "fractions".
2. Consequently, Cantor took Weierstraß "elements" to be the familiar "fractions $\frac{k_i}{n_i}$ *resp.* $\frac{l_j}{m_j}$".
3. As fractions may have either a positive or a negative sign, Cantor was able to *remove* the (ordered) pair! Because our "(a, b)" had not yet been invented, Weierstraß wrote "$\mathcal{A} + (-\mathcal{B})$" instead—and Cantor mistook Weierstraß' *formalism*.
4. However, the real historical paradox is this: Cantor did not grasp that Weierstraß used the concept of "set" to construct his "irrational" numbers! Instead, Cantor relied on the then usual concept of *ordered* set and defined a "real" number to be a *sequence* of positive or negative fractions.
5. In a complete misunderstanding of Weierstraß, Cantor finally demanded the sequences to be "convergent"—in regard to the strict deductive procedural character of Weierstraß' teaching, it is *absolutely clear* that Weierstraß was *unable* to introduce the deeply analytical notion of "convergence" into the groundworks of his theory. But Cantor did.

All told, Cantor's new concept of "real number" is this:

> A "quantity" is a *convergent* SEQUENCE of rational numbers:
>
> $$a_1, \quad a_2, \quad a_3, \quad a_4, \quad \dots$$

Cantor's definition reads this way:

> This fact I express in the following words: *"The sequence[1] has a definite limit b".*

Cantor calls "this fact" the "limit". "This fact" is a "converging sequence" (of rational numbers). In accordance with Cantor, the converging sequence (of rational numbers) is the new "real" number, the "limit"—although Cantor writes "has" instead of "is". He does not choose his words very carefully.

To emphasize again:

- Where Weierstraß thought "elements" to be multiplicities of "units" $k_i \cdot \frac{1}{n_i}$, Cantor simply assumes "rational numbers" $a_k = \frac{\alpha_k}{n_k}$.
- Weierstraß' "elements" of "aggregates", deprived of order, are turned into well-arranged "terms" of a "sequence".
- Weierstraß' formal ordering of two irrational numbers is skipped by Cantor; instead, Cantor takes recourse to *signed* numbers (rational numbers).
- Most importantly: whereas Weierstraß thought a number to be a "sum" (an *elementary* notion of *arithmetic*), Cantor implemented the *deep* idea of "convergence"—which is an *analytical* concept. Consequently, Cantor's notion *cannot* (legitimately) be introduced as a preliminary to a deductive construction of analysis.

Synopsis:

> Cantor made the following changes to the notion of "quantity":
>
> 1. The "elements" are turned from fractions without order into ordered "rational numbers".
> 2. The purely *formal* "order" (of two irrational numbers to make up one real number) is replaced by the introduction of *real* "signs"—in other words: the negative numbers are taken as an *actual* formative ingredient.
> 3. The arithmetical "sum" is upgraded to an analytical "sequence", together with its "limit".

Each of these changes replaces elementary ideas by more complicated ones. Taken together, they override all of Weierstraß' elementary properties of "quantity" and deprive his concept of "real" number of the *building blocks* for the foundations of analysis.

In 1872, Cantor invented a clear *concept* of number, detached from all *intuitive* ingredients—and technically broad enough to base wide areas of ordinary analysis on firm ground. (The "conditionally"—or "not absolutely"—convergent series were included.) Hankel was refuted—all was just great!

However: no progress without a price!

---

[1] Cantor actually used the notion "series" ("Reihe")!

The Philosophical Price of This Progress

Every major change is a matter of give and take. No progress (e.g. a technical one) comes without drawbacks (e.g. social or ecological ones). Therefore, we ask: what are the losses in regard to mathematics because of this conceptual change proclaimed by Cantor (and Heine)?

The answer to this simple question is equally simple: its *grounding* in real life. This is important, and we emphasize it:

> The acceptance of Cantor's concept of number separated analysis from reality.

Our "Brief History" has no room for such philosophical details, but nevertheless we may say: the fundamental aim of the *substancial* thinker Weierstraß is to prove that the basic concepts of his theory (which is analysis) are *abstractions* taken from the real world. He demonstrates this for his concepts of number.

As a consequence of this, his basic notions are of utmost *generality:* "set", "unit", "exact parts of the unit". Additionally, an impressive testimony to this view is the fact that the negative numbers are mere conceptual constructions, not abstractions from reality. In plain language: for Weierstraß, negative numbers do not exist.

Cantor does the opposite. Right from the start, he implements very abstract *mathematical* notions: "order", "sign" and finally the tricky concept of "convergence"! While Weierstraß toils with philosophical abstractions based on reality, Cantor performs conceptual acrobatics right from the start.

> Cantor thinks of his analysis (a) as *without any relation* to reality, (b) in *mere technical mathematical* concepts.

Cantor conceives analysis purely formally—more precisely: analysis as a whole. This arises from the fact that his basic concept (converging sequence) is a formal one.

It is not Cantor, but **Heine,** who goes one step further and provides a *purely formal ontology* for his concepts. Heine, not Cantor, takes the numbers to *be* nothing more than signs [Zeichen]. Heine writes:

> *Requirement:* To add to each number series a sign.
> One introduces the series itself as a sign, put in brackets, such that e.g. the sign belonging to the series a, b, c, d, etc. is $[a, b, c, d$, etc.$]$.

In addition, in § 2 of his article, he puts:

> *1. Definition. General number or number-sign [Zahlzeichen]* is the name of the sign belonging to the series of numbers.
> *2. Definition. Number-signs are called equal* or are interchangeable, if they belong to equal series of numbers, and *not equal* or not interchangeable, if they belong to different series of numbers.

Heine declares in crystal-clear words "numbers" to be "signs".

Cantor's contemporary Gottlob Frege (1848–1925), who is a reliable philosophical nit-picker, sharply contradicts him in 1893:

> We see that the number-signs have an entirely different importance from that which was attributed to them before the rise of formal theories. They are no longer mere external auxiliaries like blackboards and chalk; rather, they comprise the essential components of the theory itself.

On the other hand, **Cantor** argues against this plain formalism Heine was devoted to (see on p. 191).

Cantor defended himself against philosophical criticism like Frege's, in 1889, as follows:

> I never claimed that the signs b, b', b'', ... were, strictly speaking, *concrete* quantities. As *abstract objects of thought* they are merely quantities in a figurative sense. The *decisive* feature is that with the help of these abstract quantities b, b', b'', ... it is possible to determine *real, concrete* quantities exactly.

To this, **Frege** had a clear answer (which is also a criticism of Dedekind, see p. 202):

> It demands a strong faith indeed, to take signs that are written with chalk on a blackboard or with ink on a paper, that can be seen with the bodily eye, as abstract objects of thought. It is the kind of faith which moves mountains and creates the irrational numbers.

The philosophical mind Frege insists on the basic question to the new formal mathematics of Cantor, Heine, Dedekind (he will soon have his say), their disciples, and followers:

> Why have arithmetical equations a practical application? Solely, because they express thoughts. How could an equation expressing nothing, which was no more than a group of figures that could be transformed, according to certain rules, into another group of figures, be applied? Now, it is applicability alone, which elevates arithmetic from being a mere game to the rank of a science. Thus, applicability necessarily belongs to it. Is it advisable then, to exclude from arithmetic what it requires to become a science?

At the same time, Frege, with his notion of "applicability" offered the mathematicians a comfortable excuse: they could confine themselves to "pure" theory—applications were the job of the practical men, not theirs.

This was not a smart move by Frege. It is not the excuse of a division of labour between "pure" and "practical" mathematics that is at stake. Mathematics is a *whole* (which, of course, has many fields—this is one of the viewpoints of this book). At stake is the *reference to reality* of mathematics, those *components of reality* that allow the specialists to model our world thereon, as Weierstraß had been still aware of—we shall continue this line of thought at the end of this section, p. 198.

The Mathematical Price of This Progress

The concept of number that Cantor invented has two important mathematical consequences.

> 1. It is generous to such a degree that it allows for the validity of all conventional arithmetical methods.

We remember: Weierstraß judged some of the generally accepted ways of calculating to be "fallacies" (p. 177).

Oh, well: Cantor does not really allow *all* the calculations, namely not those from Johann Bernoulli and Euler, which embrace infinitely large or infinitely small "numbers" (see the example on p. 76). Because:

> 2. The infinitely small "numbers" were excluded from analysis (and therefore also the infinitely large ones).

The latter is the result of the following reasoning: Cantor learnt from his teacher Weierstraß that the definition of a new notion of number also *requires* the definition of the basic arithmetical concepts "equal", "addition", "subtraction", "multiplication" etc. *in regard to the new objects.*

The operations are simple. In the case of two converging sequences of rational numbers

$$a_1, \quad a_2, \quad a_3, \quad a_4, \quad \ldots,$$
$$a_1', \quad a_2', \quad a_3', \quad a_4', \quad \ldots,$$

one operates as follows to get

$$a_1 \circledast a_1', \quad a_2 \circledast a_2', \quad a_3 \circledast a_3', \quad a_4 \circledast a_4', \quad \ldots$$

($\circledast$ is one of the four basic arithmetical operations; of course, division by zero is forbidden) a converging sequence of rational numbers. For the "definite limits" b and b', as well as for the result of the calculation—which is the "limit" b'' of the last sequence—this means:

$$b \circledast b' = b''.$$

So far, so good.

But what about the last "="? Under which circumstances shall we take two converging sequences of rational numbers to be "equal"?

Cantor argues as follows:

> The two sequences $a_1, a_2, a_3, \ldots$ and $a_1', a_2', a_3', \ldots$ always relate to each other in one of the following three ways, each of which excludes the others: either 1. $a_n - a_n'$ becomes infinitely small for increasing n or 2., starting from a certain n, $a_n - a_n'$ remains always larger than some positive (rational) quantity ε, or 3., starting from a certain n, $a_n - a_n'$ remains always smaller than some negative (rational) quantity $-\varepsilon$.

In the first case, I establish

$$b = b',$$

in case of the second $b > b'$, in case of the third $b < b'$.

In other words, Cantor *identifies,* for example, the two new numbers

$$1, \tfrac{1}{2}, \tfrac{1}{4}, \tfrac{1}{8}, \ldots \quad \text{and} \quad 0, 0, 0, 0, \ldots$$

His argument: both numbers $1, \tfrac{1}{2}, \tfrac{1}{4}, \tfrac{1}{8}, \ldots$ and $0, 0, 0, 0, \ldots$ have the same *limit*—and "limit" is just the *name* Cantor has given to his new numbers b, b', etc.

Let us postpone the discussion as to whether this is *really* clear for a moment. Instead, we stick to Cantor's way of doing things. We realize:

> Cantor *identifies* the classical "infinitely small" variable $1, \tfrac{1}{2}, \tfrac{1}{4}, \tfrac{1}{8}, \ldots$ with the number 0.

This *and only this* makes the "infinitely small" quantities in the late nineteenth and the twentieth century disappear: because of a pure *definition*—and not because of Weierstraß and his intricate epsilontics!

As we have shown: Cantor *did* this. However: is he actually *allowed* to do this? Is such a *definition* of "equality" really permitted? It seems, after all, that the two numbers *taken to be* "equal" are *different!*

Notoriously, Mathematicians Call Different Things "Equal"

The logician Frege opposed this kind of sloppiness—seen in Cantor's reasoning— quite firmly in his later years. Frege writes:

> With respect to the equality sign, we will do well to abide by our stipulation, according to which equality is complete coincidence, identity.
>
> To be sure, bodies of equal volume are not identical; however, they have the same volume. In such a case, the signs on each side of the equality sign are not be taken as signs for bodies but for their volumes, or alternatively for the measuring numbers resulting by measurement in terms of the same unit of volume. We will not be speaking of vectors as equal but of a particular specification—let us call it "directed length"—for these vectors, which may be the same for different vectors.
>
> According to this conception the progress of science will not require an extension of the formula $a = b$; instead there are only new determinations (modi) of the objects which are to be considered.

Two decades earlier Frege's view had been this:

> Whether we say, like Leibniz, "the same" or "equal" does not matter. "The same" looks like a complete coinciding, whereas "equal" only expresses a coinciding in this or that respect; however, one can introduce such a manner of speech that this difference vanishes. For example, one says "the lengths of these straight lines are equal" instead of "these straight lines are equal in length" [...] In this manner we used the word in the examples above. Indeed, all laws of equality are covered by the general replaceability.

(Weierstraß, too, took replaceability to be the essence of equality.)

In consequence, following Frege's first opinion by calling only identical things "equal", no mathematics will result, but only logic. Mathematics usually uses "equal" in the sense given directly above. However, it is this "equal (in a certain respect)" that is often not understood by non-mathematicians, for the mathematicians usually do *not* EXACTLY *say* what they mean.

1. $x_0 = 2$ means: "The value of the indefinite x_0 is 2". So the "indefinite" is going to be "determined"!?—Well, is this permitted? Is it, actually, *permitted* to "determine" an "indefinite"? *Why* is this permitted—as it is then *no longer* "indefinite"?—Besides, it is not clear whether this is an *assumption* or a *solution* (to a question).
2. $x_0 = X$ means: "x_0 and X are the same value (without saying, which value it is)". Here two different "indefinite" values are determined *"in the same sense"* —without specifying, *what* this "determination" is going to be.
3. $f(x) = x^2$ means: "f is the quadratic function". This is an assumption or a solution. "f" is the *name* of a function.
4. $\lim\limits_{x \to 0} \frac{1}{x} = \pm\infty$ means: "It can be proved that the function $\frac{1}{x}$ has the two 'limits' $+\infty$ and $-\infty$, if $x \to 0$".
5. **etc.**

In mathematics, equality has very different meanings. It is only this *unspoken knowledge* that enables mathematics.

> Mathematics is a *speaking,* a *writing.*

The *procedure* of speaking or writing is inseparably connected to mathematics—which are both part of *making mathematics* a *creative act.* However, normally, this is not spoken of.

Weierstraß proved the following:

Theorem. *If a and b are fractions, a = b implies b = a.*

This "equality" results from a "transformation". However, this is not always the case. Another "equality" may result from an "identification" (not taken in its logical, but in its mathematical sense: identical *in a certain respect*).

An Unexplored Mathematical Potential (Of Cantor and His Contemporaries)

Let us return to the question asked on p. 195:

Is Cantor *allowed* to identify any two of his new numbers when they differ by an infinitely small quantity (to use the terminology of that time)?

Weierstraß would have demanded a *justification* for this definition. (This might be the reason for Weierstraß to *abstain* from commenting on Cantor's way of constructing the concept of the real numbers.) By contrast, Cantor did not see any necessity for this. Instead, he claims: "The *essence of mathematics* is its freedom".

There is no harm in mathematicians reflecting on their doing. Let us make up for this and add at least some thoughts.

Of course, the mathematician is *permitted* to *fix a definition*—whenever it makes sense, it is free from contradictions and is productive. It is beyond all doubt that Cantor's definitions fulfil these requirements. Nevertheless, we may add our reflections. The real question, of course, is: *must* Cantor give these definitions?

To pose this question is almost a denial of it. As nearly always in life, alternatives do exist. However, Cantor discredited them a priori by choosing the name "limit"; remaining more open, he could have spoken of a "(new) number" or something similar.

In taking Heine's formal attitude seriously, one may arrive at the following idea:

Alternative definition. Two converging sequences of rational numbers

$$a_1, \quad a_2, \quad a_3, \quad a_4, \quad \ldots,$$

$$b_1, \quad b_2, \quad b_3, \quad b_4, \quad \ldots$$

(with the same limit) are called "equal" only if they are *identical*. The equality of two corresponding terms a_n and b_n is established by the *usual* arithmetical equality.

That means: not only $1, \frac{1}{2}, \frac{1}{4}, \frac{1}{8}, \ldots$ and $0, 0, 0, 0, \ldots$ are "different", but both are also different from $1, 0, \frac{1}{4}, \frac{1}{8}, \ldots$ —Nevertheless, $1, \frac{1}{2}, \frac{1}{4}, \frac{1}{8}, \ldots$ and $\frac{2}{2}, \frac{2}{4}, \frac{2}{8}, \frac{2}{16}, \ldots$ are "equal".

It is even possible to go further, as Detlef Laugwitz showed in the 1970s. One can say: in analysis the limits do not depend on a finite number of different pairs of corresponding terms. As we are talking about the convergence of sequences of rational numbers, the above definition can be made weaker:

Alternative definition II. Two converging sequences of rational numbers

$$a_1, \quad a_2, \quad a_3, \quad a_4, \quad \ldots,$$

$$b_1, \quad b_2, \quad b_3, \quad b_4, \quad \ldots$$

(with the same limit) are called "equal" only if they are, *with the exception of only a finite number of terms*, identical. The comparison of two corresponding terms a_n and b_n is established by the *usual* arithmetical equality.

Obviously, these two changes in Cantor's definition of equality have important consequences:

1. Now there exist "infinitely small" numbers of a new kind. Take, e.g. $1, \frac{1}{2}, \frac{1}{4}, \frac{1}{8}, \ldots$ It differs from $0, 0, 0, 0, \ldots$ (and is even greater than it).
2. It is not always possible to compare two new numbers, e.g. $1, 0, -\frac{1}{3}, 0, \frac{1}{5}, 0, -\frac{1}{7}, 0, \ldots$ and $0, 0, 0, 0, \ldots$
3. Sometimes, division is problematical.

Some properties can be salvaged, if *more* numbers of this new kind are accepted.

(a) Besides the converging sequences of rational numbers, one should also accept the sequences of their reciprocals (which are the "definitely" diverging sequences of rational numbers). The reason is this: analysis demands the division of numbers. Therefore, if one accepts "infinitely small" numbers, the arithmetical laws demand "infinitely large" numbers, too. If $a \neq 0$ is a number, so must be $\frac{1}{a}$.

(b) Nevertheless, there are shortcomings that cannot be overcome: no number of this new kind, which is a sequence containing zero as a term, can be a divisor (at least in case of there being infinitely many zeros). In other words:

> This new set of numbers does not come with a general concept of division.

It is a question of personal taste whether or not one finds this acceptable. (There is an analogous state of affairs in regard to Weierstraß' numbers—however, they *can* be divided by a particular procedure! Maybe there also *exists* a *procedure* to divide by "infinitely small" numbers for this new kind of number?)

This should suffice. It was our aim to explore the possibilities of Cantor and his colleagues defining new numbers *in another way* from what they actually did. Of course, we have to concede that they did not see these possibilities. It was not until the second half of the twentieth century that such ideas have emerged.

The Formal Character of Cantor's Analysis

Frege's criticism (p. 193) made it clear: Cantor's analysis bears a formal character—although not to the same degree as Heine's. The reason is: *Cantor is unable to declare* WHAT *his numbers b, b', ... are.* To Frege, Cantor replies: they are *things of thought.* However, in 1872, Cantor was unable to specify, *of what particular kind* these "things of thought" are. Whereas his teacher Weierstraß tried to create mathematical concepts fit for *real applications*, Cantor was content with "things of thought". Not Cantor, but Weierstraß justified the introduction of *each* new concept of number with recourse to real activities .

It took another 23 years before Cantor succeeded in proposing a first draft of what is called "set-theory" nowadays. The first sentence of his treatise from 1895 is widely known:

> By "set" (in German: "Menge") we mean any collection M of definite, clearly distinguished objects m of our intuition or our thinking (which will be called "elements" of M) of a whole.
> It will be symbolized thus:

$$M = \{m\}.$$

There had been earlier attempts, but not until Cantor's ideas did they lead to a system. During the twentieth century, these ideas developed gradually into "set-theory", or more exactly: into *different* set-theories.

> Set-theory is a theory of pure thought, without any chains to reality (and even without any postulation thereof).

Ever since Cantor published his first sketch of a *Set-theory*, the real numbers have been constructed *within a set-theory*. (Other attempts did not gain wide recognition.) That is why it is legitimate to characterize Cantor's concept of number as a "formal" one—a concept that does *not want* to be related to *anything* approaching reality.

This is a decisive change within the philosophical style of analysis. This change happened between Weierstraß and Cantor, and it was driven further by the creation and the implementation of Topology into the field of analysis, which is still the case today.

The Form of the Real Numbers, Invented by Traugott Müller, Joseph Bertrand, and Richard Dedekind

The Basic Problem

On 20 March 1872, Richard Dedekind (1831–1916) wrote the following:

> The statement is so frequently made that the differential calculus deals with continuous magnitude, and yet an explanation of this continuity is nowhere given. [...] To secure a real definition of the essence of continuity I succeeded on 24 November 1858. [...] I could not make up my mind to its publication, because the theory itself had little promise.

However, on March 14 and 20, the two treatises by Heine and Cantor with their construction of the real numbers came into Dedekind's hands, and this made him publish his own ideas in a *commemorative* that appeared on 26 April 1872.

Subsequently, the text was written in a hurry. This is readily apparent. Later, Bertrand Russell presented the idea much more elegantly.

Even his choice of words turned out to be somewhat clumsy. He called "stetig" ("continuous") what is normally called in German "kontinuierlich" (in English they are the same! see p. 108).

The Basic Idea

Clear-sightedly and from a historical distance: what was the actual problem? What was *really and "securely"* known about the "numbers" in 1872?

Well, with regard to analysis, nobody will doubt the nature of the "natural" numbers 1, 2, 3, ... Similarly with the "fractions" ($\frac{1}{2}$, $\frac{1}{3}$, $\frac{2}{3}$, $\frac{3}{7}$, ...) that had been used in trade and commerce for centuries.

Really problematic were (and are!) the other "numbers": "negatives", "roots" ($\sqrt{2}$, $\sqrt[3]{5}$, ...), or even more extravagant numbers like, for instance pi (π) etc. What was known about them?

At least, this: they can be indefinitely approximated by the unproblematical numbers (the fractions). The problematical numbers are placed, somewhat cosily, *in between* the unproblematical ones. We have already looked at this (p. 168).

Consequently, all things considered, the following idea will appear not too far-fetched:

Let us classify the fractions according to whether they are *smaller* or *greater* than such a problematic number!

The most important thing with this idea, as with any good idea is: *one must* HAVE *it!* Nothing else can get off the ground without this *flash of genius*. However, for those who succeed, everything will follow.

The idea is this: *define* such a problematic number *by classifying* the known numbers—into whether they are *smaller* or whether they are *greater* than the one under consideration. Isn't this idea much easier than Cantor's—isn't it closer to hand?

Yes, it is! This is proved by the fact that it came up much earlier than Cantor's—for: it is a fact, Richard Dedekind was not the first who thought of it.

Traugott Müller, the Very First Man Who Invented the "Dedekind-Cuts", in 1838

A Mathematical Outsider

Traugott Müller (1797–1862) was a very gifted mathematician. His father was a priest, and the son studied first theology and then mathematics and science in Leipzig between 1817–22. In the early years, when Müller was gradually introduced in traditional German grammar schools (called *"Gymnasien"*)—in which, originally, only languages were taught—, Müller worked as the headmaster at the *Real-Gymnasien* of Gotha (1836–44) and Wiesbaden (1844–62). Besides other books, he published a *Textbook of General Arithmetic* for grammar schools, comprising 555 pages, in 1838; a second edition appeared in 1855.

Müller's Educational Convictions

Müller's aim was to make Arithmetic as important as Geometry in education. Both fields "should keep the mind engaged to an equal extent". This aim led Müller to write his highly systematized textbook of Arithmetic, which he started with a systematic classification of the concept of number.

Müller's Concept of Irrational Number

Having introduced the *fractions* (without sign), Müller presented the classical proof for the existence of "incommensurable" quantities. From this he concluded:

> Since, for two incommensurable quantities A and B, we always have
>
> $$\frac{p}{q} < \frac{A}{B} < \frac{p+1}{q}.$$
>
> Whatever the integer q might be, we are not able to represent the quotient $\frac{A}{B}$ exactly, neither by an integer nor by a fraction. However, we can imagine q as large as we wish and, consequently, make $\frac{1}{q}$ smaller than any given number. That is why it is possible to approximate this quotient $\frac{A}{B}$ endlessly, whereby $\frac{A}{B} - \frac{p}{q}$ as well as $\frac{p+1}{q} - \frac{A}{B}$ is always less than $\frac{1}{q}$. This makes $\frac{A}{B}$ the *limit* of these two fractions, which will change for any new q.

And then follows a breathtaking (somewhat long-winded) sentence:

> If we would be able to show that these limits, if they were joined by the usual operation signs, can be transformed in the same way as were proved for the integers and later for the fractions, then it would be legitimate to give them the name *number* with the same good reasons as was done with fractions, thereby raising the former theorems to more generality. This will be done in the following paragraphs.

Müller continues:

> Unlike the former numbers, these limits will be called *irrational numbers*, whereas the former ones are named *rational numbers*.

Next, Müller proves the equality of his irrational numbers to be the identity and shows the validity of the commutative law for addition and multiplication for them. However, Müller's definition of the relation $<$ is wrong.

Nevertheless, we can state:

(a) In 1838, Traugott Müller formulated the construction, which today bears the name of Dedekind ("Dedekind-cut"), in absolute clarity.
(b) Müller took this idea to define a new kind of number, beyond the rational numbers.

An Incidental Idea, by Bertrand in 1849

Eleven years later, a textbook on arithmetics for scientists was published in Paris. Its author was Joseph Louis François Bertrand (1822–1900), who was, like Müller, a grammar schoolteacher but became in 1856 a university professor. It is almost certain that he was unaware of Müller's book, but, amazingly, he also formulated Müller's idea:

> An irrational number cannot be defined by means of a unity, like a quantity. That is why we take for its definition *the specifications, which rational numbers are smaller* and *which are greater than* it ...

... which is the irrational number. Quite casually and without pretentiousness, all necessary aspects are stated here.

The logician might object: how can something known to be "smaller" or "greater" than something else, which is unknown? However, this is nit-picking. Bertrand presents the idea *very* clearly: he *divides* the "rational" numbers into two classes, the class of the "smaller" and that of the "larger" ones.

Spelling out all the details remained a task for Dedekind 13 years later. But even so, the idea was formulated.

Bertrand continued in this casual style. *Addition* is left out—what could possibly not be clear about it? However, *multiplication* is treated more explicitly:

> If the multiplicator is irrational, we need a new definition [of multiplication]. We call the product of a [rational] number A with an irrational number B a number, which is smaller than the product of A with any arbitrarily chosen rational number, which is greater than B and which is greater than the product of A with any arbitrarily chosen rational number, which is smaller than B.

The product of a rational number r with an irrational number q *cannot* be rational. And how shall we determine this irrational product? Clearly, by specifying the "smaller" and the "greater" rational numbers. All this is stated by Bertrand.

All well explained. However, Dedekind presents just the same, but with more pomposity.

The Dramatic Version, Given by Dedekind in 1872

Richard Dedekind gives a much more dramatic presentation of this simple idea— even without considering his geometrical preliminaries.

First, Dedekind defines the concept of "cut" (which nowadays bears his name; by "system R", Dedekind means all the rational numbers):

> If there is given any separation of the system R of rational numbers into two classes such that every number a_1 of the first class A_1 is less than every number a_2 of the second class A_2 [...], then for brevity we shall call such a separation a *cut* and designate it by (A_1, A_2).

This is the crystal-clear definition, which was also given, and in nearly the same precision, by Traugott Müller in 1838, and, somewhat more incidentally, in 1849, by François Bertrand.—What follows is Dedekind's description of such a *problematical* number:

> Whenever, then, we have to do with a cut (A_1, A_2) produced by no rational number, we create a new, *irrational,* number α, which we regard as completely defined by this cut (A_1, A_2). From now on, and for the sake of brevity, we shall say that the number α corresponds to that cut or that it creates this cut.

What Müller declares to be a "legitimate" number and what Bertrand plainly *defines* to be a number is highly stylized by Dedekind to be a "creation". It's the tone that makes the music. Frege's response we do already know (p. 193).

The arithmetical operations are given by Dedekind this way:

> If c is any rational number, we put it into the class C_1, provided there are two numbers one a_1 in A_1 and one b_1 in B_1 such that their sum $a_1 + b_1 \geq c$; all other rational numbers shall be put into the class C_2.
>
> This separation of all rational numbers into the two classes C_1, C_2 evidently forms a cut, since every number c_1 in C_1 is less than every number c_2 in C_2.

These are, yet again, the same ideas that Müller and Bertrand gave but garnished with a lot of pomp and in much more detail.

The Elegant Version, Given by Russell, in 1919

Half a century later, Russell (1872–1970) put things in a rather elegant and simple way. (The Nobel laureate for the literature, Bertrand Russell, wrote this book during his six month imprisonment for his militant anti-militarism.) Felix Hausdorff (1868–1942) got the same idea earlier but did not explicate it with the same elegance.

(i) Russell started with the concept of dividing the rational numbers into two classes, where the class of the smaller numbers has no maximum and the class of the bigger one no minimum. (ii) Then he defined the concept of "limit" (to be "of utmost importance"): the "upper limit" in the case that the class of smaller numbers has a maximum and the "lower limit" in the opposite case. Thereafter, he proceeded as follows (*emphasized* words added):

> Let us observe that an irrational is represented by an irrational cut and a cut is represented by its lower section.
>
> *Definition (segment)*. Let us confine ourselves to cuts in which the lower section has no maximum; in this case we will call the lower section a "segment".
>
> Then those segments that correspond to ratios are those that consist of all ratios less than the ratio they correspond to, which is their *boundary*[2], while those that represent irrationals are those that have no boundary.
>
> Segments, both those that have boundaries and those that do not, are such that, of any two pertaining to one series[3], one must be part of the other; hence, they can all be arranged in a series by the relation of whole and part, *that is the relation* $\subseteq$.
>
> A series in which there are Dedekind gaps, i.e., in which there are segments that have no boundary, will give rise to more segments than it has terms, since each term will define a segment having that term for boundary, and then the segments without boundaries will be extra.
>
> We are now in a position to define a real number and an irrational number.
>
> *Definition (real number)*. A "real number" is a segment of the series of ratios in order of magnitude.
>
> *Definition (irrational number)*. An "irrational number" is a segment of the series of ratios which has no boundary.
>
> *Definition (rational real number)*. A "rational real number" is a segment of the series of ratios which has a boundary.

---

[2] emphasis added

[3] The modern term is: "sequence".

Thus a rational real number consists of all ratios less than a certain ratio, and it is the rational real number corresponding to that ratio. The real number 1, for instance, is the class of all proper fractions [all negatives included].

Russell simplifies Dedekind's concepts significantly:

(i) Russell realizes: Dedekind does not really work with *two* classes A_1, A_2 of rational numbers, but only with one: A_1; for A_2 is defined by A_1: $A_2 = R \setminus A_1$ (the set-theoretical difference).

(ii) Whether A_1 has a boundary or not is important. (This boundary may belong to A_1 or to $R \setminus A_1$.)

(iii) Consequently, Dedekind's use of the ordered pair (A_1, A_2) (at least: in notation) is empty talk. Actually, Dedekind does not work with pairs (ordered or not), but only with single sets of rational numbers (of a certain kind).

This is the fourth construction of the "real" numbers, which makes use of the rational numbers or fractions. That this idea was published three times, probably independently of each other, in the middle of the nineteenth century, seems to show at least two facts:

(a) Textbooks for schools were not read by scientists in those times.

(b) Around 1872, German mathematicians were not used to reading French text books—which was obviously an omission.

However, the first point might not be completely correct. For Cantor cited a schoolbook in 1883! And even more surprisingly: he cited Müller's textbook for grammar schools in its second edition! He did this in his introduction, where he mentioned the rational numbers, in which Cantor continued to discuss the concepts of real numbers ("which are known to me"), given by Weierstraß, Dedekind, and himself.

We know already that Cantor misunderstood Weierstraß' concept, but what about Müller's? The second edition of Müller's book also contains his concept of real number, which at bottom is the same as Dedekind's. This is true, although in the second edition it is not as clearly presented as in the first edition. Now: did Cantor not understand Müller's definition of the real numbers? (If so, this would strengthen the suspicion that mathematicians are not reliable readers of alien ideas.) Or did he not want to discredit Dedekind's fame of being the (first and sole) inventor of this concept?

The Evaluation of This Solution by Tannery in 1886

Because nobody knew about Weierstraß' special concept of real number in the nineteenth (and twentieth!) century (there may have been one single exception: Otto Hölder (1859–1937)), only two of them became public knowledge. But which of them was thought to be the proper one for the foundations of analysis?

In 1886, Jules Tannery (1848–1910) published his textbook *Introduction to the Theory of Functions in One Variable*, which became influential in the following years. Therein he defines the real numbers with reference to Bertrand. Later, Tannery explained that he had not had access to Dedekind's treatise from 1872, but he had known about Heine' detailed treatise, that is: Cantor's concept.

Tannery's evaluation is of interest: he valued the constructions of Cantor and Heine to be "too arbitrary", and therefore he did not accept them. A very interesting judgement! According to Tannery, the construction of the real numbers as "cuts" within the class of the rational numbers is the more natural way, free of arbitrariness. Tannery accepted the fact that the operations using "cuts" are far less elegant than those with sequences—not to speak of the proofs of the arithmetical laws! Today we may guess: if Tannery had known about Weierstraß' construction of the real numbers, he might have preferred it to the others because of the better coherence of the formation of its concepts, the simplicity of its means, and the more general way of defining addition and multiplication. Instead, there might have arisen a fruitful discussion about the *real differences* between direct and indirect arithmetical operations. Perhaps, analysis would have taken on a different shape?

The Axiomatic Characterization of the Real Numbers by David Hilbert in the Years 1899 and 1900

The Situation in 1872: Two Definitions of One Subject

Müller as well as Bertrand and Dedekind on the one side, and Cantor on the other, had published *two* constructions of the real numbers. Although they differ significantly from each other, both constructions were based on the "fractions".

> Why do both constructions lead to the *same* result? Moreover, do they *actually* produce the same result? And if so, what exactly is the meaning of "the same"?

So far as I am aware, this question was not formulated in print before Hilbert's Axioms. Clearly, the fact is *not evident* and a proof for this sameness is far from evident. To my knowledge, no mathematician had any doubts then. The *fundamental belief* in the *unity of mathematics* might have been the underlying sentiment ruling out uncertainty.

Hilbert's Axiomatic System—the First Attempt, 1899

Without having been posed beforehand, the problem was solved 27 years later by David Hilbert (1862–1943). He presented a *characterization* of the "real" numbers.

Hilbert approached the problem differently from earlier mathematicians. He argued for a *logical* perspective. As for Geometry, Hilbert defined the task for mathematics to offer "a logical analysis of our spatial intuition". In the process of this, he presented in 1899, *quite casually,* a logical examination of the analytical concept of number. Hilbert wrote:

> In the beginning of this chapter we are going to give a short discussion of complex number systems, which will be helpful later on in the presentation.
> The real numbers in their entirety are a system of things having the following properties:
> . . .

Hilbert's wording "complex number system" shows the influence of Weierstraß. However, what follows is completely alien to Weierstraß' way of thought:

twelve "theorems of operation"

(like: $a \cdot 1 = a$ or $a(bc) = (ab)c$)

as well as four "theorems of order"

(e.g.: if $a > b$, then we have $a + c > b + c$)

and the "Theorem of Archimedes".

That was all!

Or was it not, after all? Immediately after his book was published, Hilbert realized two mistakes. First, he omitted a very important "property"—we shall come to this soon. Secondly, in his first characterization of the system of real numbers, Hilbert called the properties under discussion "theorems". However, a "theorem" in mathematics is a proposition that must be proved. Nonetheless, Hilbert *is unable* to prove these twelve "theorems"! How could he? Hilbert has *nothing at all* at hand for what he may *prove* "properties" thereof! Quite contrarily, Hilbert takes the opposite way. Hilbert *states* some "properties"—and then proclaims: I shall call all objects that bear these properties to be the "system of real numbers". Without any argument: there you go! Par ordre du mufti.

At the first attempt, Hilbert made two mistakes: a *methodological* one (his talking about "theorems") as well as one of skill (the omission of a central property). Even the great mathematician David Hilbert was not infallible.

The Axiomatic System of Hilbert—the Second Attempt in 1900

One year on, Hilbert published an article, wherein he removes those defects (without hinting at these changes).

In the new paper, Hilbert repeats his theorems word for word—but calls them "axioms". Only their order is, in two cases, changed. Moreover, Hilbert changes and supplements titles. Now there are six "axioms of operation", six "axioms of calculation", as well as four "axioms of order". Also, there are two "axioms of continuity" introduced: as the last "theorem" ("Theorem of Archimedes", now: "Axiom of Archimedes"), and as eighteenth axiom (*emphases* are added):

> *Axiom of Completeness.* It is not possible to add to the system of numbers another system of objects such that the preceding axioms are fulfilled for the resulting new system. In short:

the numbers constitute a system of objects which is not capable of any enlargement, if all the axioms hold.

This decides the question that divided the two collaborating colleagues Cantor and Heine and in which Dedekind sided with Heine's position by a pure postulation. Hilbert decided that Dedekind and Heine were right and that Cantor was wrong.

The problem was the following. Cantor as well as Heine *generated* from the system of rational numbers a new system of numbers (today called: "real" numbers)—although with different methods.

In both cases, the question arose: what happens, if the new system, i.e. the real numbers, is taken as a starting point, instead of the "rational" numbers, and if we then *generate in the same manner* yet another system of new numbers?

Heine formulated and proved a theorem: the numbers that are subsequently generated "are not new, but coincide with those which were generated at first". (That this theorem contradicts Heine's opinion that numbers were signs, as a matter of course, went unnoticed.)

Ironically, Cantor disagreed. It is true, he conceded that the two newly generated "systems of number to some extend coincide", but he wanted "to cling to the conceptual differences of the two systems". In this case, Cantor's manner of forming concepts was very pedantic.

Dedekind, on the other side, felt himself unable "to *yet* detect the advantage of this mere [*sic*] conceptual difference" (*emphases* added).

Now Hilbert decrees: this *conceptual* difference, which was "only" noticed by Cantor, is, when *considered purely mathematically,* insignificant. Basta!

Today, the two "axioms of continuity" are replaced by a single one, which is designed very differently: the "Axiom of Completeness". The "Theorem of Archimedes" is then proved (with the help of the unlimitedness of the natural numbers).

Advantage and Disadvantage of Hilbert's Axiomatic System

With his axioms, Hilbert distilled out those properties of the "real" numbers that have been generally accepted ever since. (Maybe not *completely* apt, as it seems today.) This was the first decisive step on the path to *structural mathematics,* which dominated analysis world-wide after the Second World War, and which was disseminated by the fictitious author called "Nicolas Bourbaki".

The power of this perspective is obvious: the *important* properties (of the number system) are clearly expressed. This helps to study the *variants* of the number system and their special attributes. The same with other *structures.*

However, Weierstraß' construction of the real numbers proves that Hilbert's method is not universally applicable! Weierstraß' system has no *concept* of division—nonetheless, it is *possible* to divide these numbers! Weierstraß' construction unmistakably proves: an indirect operation is *fundamentally* different from a direct one. If you want to divide two numbers, you *need* a common system to represent them (e.g. the decimal system)—otherwise you will not be able to describe a procedure *formally* (which is one of trial and error!) in order to execute the calculation! That is why Weierstraß' construction has proved that:

> There exists a version of the real numbers, which does not comply with Hilbert's axiomatic system.

This fact seems to narrow the scope of Hilbert's axiomatical method and to point to a system missed out by structural mathematics. Maybe "pure" mathematics in the footsteps of Hilbert, that is to say: structural mathematics, is not at all as *general* as it is claimed to be? Is there some pure mathematics which is not *structural mathematics?* And if so, what is it?

Hilbert did not stop at just listing these advantages. He went on to *demand:* the "real" numbers *must* be defined in this axiomatic way.

That's a bit much! Immediately, Gottlob Frege, whom we already know as a harsh philosophical critic, entered the scene. In a short correspondence between 27 December 1899 and 22 September 1900, both exchanged their opinions without reaching any consensus.

Frege asked Hilbert about the *justifications* for his axioms. Taken from a traditional philosophical point of view (which we set out to call "substancial"), *axioms* must express "the basic facts of intuition", and therefore they are in need of a legitimation.

Hilbert firmly rejected such a demand (and thereby, as a consequence, the traditional philosophical view). Hilbert said: I merely baptize certain properties (Hilbert calls them "characteristics") to be "axioms". This was merely a "matter of taste". In his opinion, a *concept* cannot be derived from a *thing*, but *concepts* could only be specified by "their interrelatedness"—as it is done by an axiomatic system. On December 29th, 1899, Hilbert explained this by an illustrative formulation:

> If I take my points to be any system of things, e.g. the system: love, law, chimney-sweeps, and if I only assume all my axioms to be the relations between these things, my theorems would be valid for these things, e.g. the Theorem of Pythagoras.

Frege opposed this and objected to it (thereby referring to *different kinds* of geometry; in some of them the Theorem of Pythagoras is valid, and in others it is not, e.g. in the spherical geometry):

> Only the wording of the theorem is the same; the mental content (Gedankeninhalt) is in each geometry another. It would not be correct to call a special case of the Theorem of Pythagoras "the Pythagoras"; because if we have proven it in a special case, we have not proven *the Pythagoras.*

The conception of mathematics, propagated by Hilbert against that of Frege, is called by me "relational". It is opposed to the "substancial" conception, which was indicated by Weierstraß. Therefore, we should not be surprised that Hilbert's method of judgement does not apply to Weierstraß' concepts. It is really a pity that Frege did not know about Weierstraß' concept of numbers—because he could *seriously* have driven him into a corner.

The New Social Duty of Mathematics—Concerning Hilbert's Ideas

In a public speech, given in Zürich in 1917 during the Great War, Hilbert resumed this topic (*emphases* original):

> If we compile the facts of a certain, more or less extensive field of knowledge, we easily realize that these facts are capable of being arranged. This arrangement always takes place with the help of a certain *framework of concepts,* such that each single object corresponds to a concept of this framework and each single fact of this field corresponds to a logical relation between these concepts. This framework of concepts is nothing else than the *theory* of this field of knowledge.

This way, Hilbert relocates traditional mathematics in a fundamentally new way in our society: the subjects of mathematics are no longer, as before, the facts (of thinking), but it is the *ordering* of any facts within any "field of knowledge", whatsoever.

> In doing so, he elevated mathematics, towards the end of the great upheaval of civilization, the Great War, to the general security police of science. From now on, mathematics did not have to deal with "things" and their "relations", but with the appropriate "framework of concepts": for the logical structuring of "every fact in a given field of knowledge". This dismisses traditional mathematics as a *science of things (concepts).*

By the way, it was Dedekind who had already demanded this basic change within mathematics more than half a century earlier. As early as 1854, Dedekind stated in his habilitation lecture:

> This twisting and turning of the definitions for the sake of the found laws or truths, in which they play a role, constitutes the highest skill of the systematist.

Dedekind did not say why "truths" are "laws". He also concealed *from where* these "laws" or "truths" should derive their authority. It seems clear that Dedekind takes the "concepts" to make up that "framework" talked about by Hilbert in his letter to Frege in 1899 and in his public speech in 1917.

The Price of Success: The Inclusion of the Actual Infinite in Mathematics

The Plain Fact

By the year 1872, following Dedekind, the idea of *defining* the "irrational" numbers with the help of "cuts", which are divisions of the ENTIRETY of the rational numbers, was increasingly accepted in mathematics. This had not been the case in Bertrand's time, in 1849, and even less so in Müller's time, in 1838.

A "cut" divides the *entirety R* of the "rational" numbers into two parts A_1, A_2, and each (!) of these parts is *infinite*, i.e. it comprises *more elements than can be expressed by any natural number.*

Each of the parts A_1, A_2 is the "complement" of the other, i.e. it contains *all* elements which the other part *does not* contain: $A_2 = R \setminus A_1$ and $A_1 = R \setminus A_2$. Consequently, each part is completely determined by the other one, and it is enough to take *one* of these two parts for the definition of the "irrational" number. This had been realized by Russell (p. 203): take a "segment" instead of a "cut", A_1 instead of $(A_1, A_2) = (A_1, R \setminus A_1)$. However, less is not enough!

> The definition of a single "irrational" number needs an infinitely large set of "rational" numbers.

An enormous expenditure! However, also the other two constructions of the irrational or the real numbers (from Weierstraß and from Cantor) are unable to dispense with infinite number sets. Maybe there is no other way?

Everybody who accepts one of these constructions as a *definition* admits to the existence of the actual infinite in mathematics. That means: she or he acts as if the *infinite* set of "rational" numbers under consideration—Russell's "segment", Cantor's "sequence" or Weierstraß' "irrational" number—was a *well-defined* object.

The State of Affairs Until Now

In Chap. 5 it was shown how much Leibniz resisted accepting the actual infinite in mathematics.

The two subsequent chapters demonstrated that Johann Bernoulli as well as Euler took the completely opposite view and *calculated* using actual infinite quantities. Euler, especially, presented highly impressive examples for the productive use of the actual infinite in his "Algebraic Analysis" (pp. 77f).

The remodelling of the "Algebraic Analysis" into the "Analysis of Values" since 1817/21 rendered the reputation of the actual infinite in analysis precarious. The reason was that as soon as "value"—and, consequently, "number"—becomes basic notions of analysis, they needed a characterization that is suitable for proofs. It must become possible to create "numbers" in proofs—which means to create them conceptually.

However, a concept of "number", which is appropriate for proofs, existed (before 1872 *resp.* 1849 or 1838) only in a *technical* sense: the decimal number.

In the beginning of the thirteenth century, the Hindu–Arabic positional system was brought into commercial Italy by Leonardo of Pisa ($c1170$–$c1250$). By the end of the fifteenth century, this technique of writing numbers had reached the trading towns in Germany and was thereby slowly superseding the Roman numerals. The Rechenmeister taught the computation "by feathers or chalk in numerals" (Adam Ries (1492–1559); that is written computation) to the businessmen as an alternative to computation on the abacus. In this process, the signs "+" and "−" started to replace the Italian abbreviations "p" and "m". Christoff Rudolff (c1499–c1545) was the first to use the decimal comma in his textbook *Hurried and Charming Computation With the Help of the Ornate Rules of Algebra, Which Ordinarily are Called the Coss*, which was printed in Strasbourg in 1525. In 1585, the Dutch businessman, engineer and physicist Simon Stevin (1548/9–1620) systematically taught the computation with decimal numbers in his little book *De Thiende* ("About the Tenth").

La Géometrie from Descartes actually shows only rational numbers. But Leibniz could not get along with them and used decimal numbers in his calculations. In 1821, Cauchy proved the Intermediate Value Theorem by dividing the distance h of the two limits x_0 and X of the domain into m equal parts $x_0 + \frac{h}{m}$, $x_0 + 2\frac{h}{m}$, $x_0 + 3\frac{h}{m}$, ..., $X - \frac{h}{m}$; the newly found subinterval $X' - x_1 = \frac{h}{m}$ is divided yet again into intervals of length $\frac{1}{m}(X' - x_1) = \frac{1}{m^2}(X - x_0) = \frac{h}{m^2}$, etc. This can be understood as the way to determine the decimal places of the required root X'.

However, nobody actually proposed a figurative representation for "infinite" numbers.

A first proposal for this only appeared in the 1950s (see p. 227).

And we have been aware (p. 177), Weierstraß expressly wanted to exclude infinity (∞) from analysis, although it was commonly accepted as being a "value".

The Compromise

In this challenging situation—the urgency of creating a *concept* of an analytical "number" that can be used in proofs, and at the same time a great scepticism towards nearly every kind of the infinite—the analysts (and even Weierstraß) were finally ready to swallow the bitter pill of accepting the *concept* of the actual infinite in 1872. By this year, the majority was ready to make a small concession and accepted the actual infinite at least *conceptually* (one or even two infinite sets of rational numbers make up a real number). However, at the same time, they decisively *insisted* on excluding the infinite *numbers* from analysis.

It is another paradox within the history of analysis that it was Georg Cantor of all people, the man who had been committed to the recognition of conceptualizing the actual infinite on a large scale, who found himself with both the obligation to *prove* and the capability of *proving* that "the existence of actual infinitely *small* quantities is impossible". Cantor, being himself one of the greatest advocates for accepting the actual infinite as a mathematical *concept*, especially in set-theory, nevertheless struggled with all his might to exclude the actual infinite from arithmetic and *actual calculations*.

Why should this be? Was it not Euler who had already shown what extraordinary calculations could be carried out with infinite "numbers" *within analysis* (see p. 77)?

One should not underestimate the influence of Weierstraß' *personal* aversion towards the inclusion of the infinite into analysis—yet, even outside of his sphere of influence (maybe in France) I don't know of any attempt to establish the infinite "as a number for calculation". It wasn't until 1976 that a mathematician (Detlef Laugwitz) dared to propagate this in print. Yet, this happened a full two decades after he had formulated an alternative version of analysis that provided for this.

Literature

Bertrand, J. (1849). *Traité d'Arithmétique*. Paris: Librairie de L. Hachette et Cie..

Cantor, G. (1872). Über die Ausdehnung eines Satzes aus der Theorie der trigonometrischen Reihen. *Mathematische Annalen, 5*, 123–132. cited from Zermelo 1932, pp. 92–102.

Cantor, G. (1887/1888). Mitteilungen zur Lehre vom Transfiniten (2 parts). *Zeitschrift für Philosophie und philosophische Kritik, 91, 92*, 81–125, 240–265. cited from Zermelo 1932, pp. 378–439.

Cantor, G. (1889). Bemerkung mit Bezug auf den Aufsatz: Zur Weierstraß-Cantorschen Theorie der Irrationalzahlen. *Mathematische Annalen, 33*, 476. cited from Zermelo 1932, p. 114.

Dedekind, R. (1901). *Essays in the theory of numbers*. (Chicago: The Open Court Publishing Company), authorised translation by Wooster Woodruff Beman. https://www.gutenberg.org/2/1/0/1/21016/.

Dedekind, R. (1854). Über die Einführung neuer Funktionen in die Mathematik. In R. Fricke, E. Noether, & Ø. Ore (Eds.), 1930–32 (vol. 3, pp. 428–438).

Dedekind, R. (1872). *Stetigkeit und irrationale Zahlen*. Braunschweig: Vieweg. see also: Fricke, Noether & Ø. Ore (Eds.), 1930–32 (vol. 3, pp. 315–334).

Frege, G. (1884). *Die Grundlagen der Arithmetik*. Breslau: Verlag von Wilhelm Koebner. reprints Breslau 1934, Hildesheim 1961.

Frege, G. (1893, 1903). *Grundgesetze der Arithmetik: Begriffsschriftlich abgeleitet* (2 vols). Jena: Pohle.   https://gdz.sub.uni-goettingen.de/dms/load/img/?PPN=PPN593233409&DMDID=DMDLOG_0001;http://gdz.sub.uni-goettingen.de/dms/load/img/?PPN=PPN593233549&DMDID=DMDLOG_0001.

Fricke, R., Noether, E., & Ore, Ø. (Eds.) (1930–32). *Richard Dedekind. Gesammelte mathematische Werke*. (3 vols) Braunschweig: Vieweg.

Gabriel, G., Hermes, H., Kambartel, F., Thiel, C., & Veraart, A. (Eds.) (1976). *Gottlob Frege: Nachgelassene Schriften und wissenschaftlicher Briefwechsel*. (Vol. 2). Hamburg: Felix Meiner Verlag.

Gabriel, G., Kambartel, F., & Thiel, C. (Eds.) (1980). *Gottlob Freges Briefwechsel mit D. Hilbert, E. Husserl, B. Russell, sowie ausgewählte Einzelbriefe Freges*. No 321 of the *Philosophische Bibliothek*. Hamburg: Felix Meiner Verlag.

Gericke, H. & Vogel, K. (1965). *Simon Stevin: De Thiende (Die Dezimalbruchrechnung)*. Frankfurt am Main: Akademische Verlagsgesellschaft.

Hankel, H. (1867). *Vorlesungen über die complexen Zahlen und ihre Functionen in zwei Theilen. I. Theil. Theorie der complexen Zahlsysteme, insbesondere der gemeinen imaginären Zahlen und der Hamilton'schen Quaternionen nebst ihrer geometrischen*. Leipzig: Leopold Voss. https://www.urn:nbn:de:bvb:12-bsb10081922-7.

Hausdorff, F. (1914). *Grundzüge der Mengenlehre*. reprint New York: Chelsea Publishing Company, 1949.

Heine, E. (1872). Die Elemente der Functionenlehre. *Journal für die reine und angewandte Mathematik, 74*, 172–188.

Hilbert, D. (1899). Grundlagen der Geometrie. In Fest-Comitee (Ed.). *Festschrift zur Feier der Enthüllung des Gauß-Weber-Denkmals in Göttingen*. Leipzig: Teubner.

Hilbert, D. (1900). Über den Zahlbegriff. *Jahresbericht der Deutschen Mathematikervereinigung, 8*, 180–184.

Hilbert, D. (1917). *Axiomatisches Denken.* cited from Hilbert 1932–35, (Vol. 3, pp. 146–156).

Hilbert, D. (1932–35). *Gesammelte Abhandlungen,* 3 vols. reprint: New York: Chelsea Publishing, 1965.

Laugwitz, D. (1976). Unendlich als Rechenzahl. *Der Mathematikunterricht, 22*(5), 101–117.

Laugwitz, D. (1978). *Infinitesimalkalkül: Kontinuum und Zahlen. Eine elementare Einführung in die Nichtstandardanalysis.* Mannheim, Wien, Zürich: Bibliographisches Institut.

Müller, J. H. T. (1838). *Lehrbuch der allgemeinen Arithmetik für Gymnasien und Realschulen, nebst vielen Übungsaufgaben und Excursen.* Halle: Verlag der Buchhandlung des Waisenhauses. https://sammlungen.ub.uni-frankfurt.de/urn/urn:nbn:de:hebis:30:4-103650.

Müller, J. H. T. (1855). *Lehrbuch der allgemeinen Arithmetik für höhere Lehranstalten. Erstes Heft. Second revised edition.* Halle: Verlag der Buchhandlung des Waisenhauses.

Ries, A. (1522). *Rechenbuch auff Linien vnd Ziphren ....* Frankfurt: Chr. Egen. Erben, 1574. reprint: Darmstadt: Hoppenstedt

Rudolff, C. (1525). *Behennd vnnd Hübsch Rechnung durch die kunstreichen regeln Algebre / so gemeincklich die Coß geneñt werden ...* Argentorati [= Straßburg]: Vuolfius Cephaleus.

Russell, B. (1923). *Bertrand Russell (1919): Einführung in die mathematische Philosophie.* In *Philosophische Bibliothek* (Vol. 536). Hamburg: Felix Meiner Verlag, 2002. German: Emil Julius Gumbel.

Tannery, J. (1886). *Introduction à la théorie des fonctions d'une variable.* Paris: A. Hermann.

Zermelo, E. (Ed.) (1932). *Georg Cantor. Gesammelte Abhandlungen mathematischen und philosophischen Inhalts.* Hildesheim: Georg Olms Verlagsbuchhandlung, reprint 1966.

Chapter 14
Analysis with or Without Paradoxes?

In this chapter, we shall introduce attempts to base the calculus on a different concept from that of Weierstraß.

Built on Very Thin Ice: Cantor's Diagonal Argument

The Presentation of Evidence

It is quite astonishing to discover which precarious arguments are sometimes accepted as being mathematically sound.

An outstanding example thereof is the (time and time again, by experts and non-experts alike) celebrated "proof for the existence of the uncountable infinity". This bizarre "proof" runs as follows (there should be added some technical details in case the number ends with infinitely many nines, but this is just technical and does not touch the essential reasoning; therefore, it can be omitted):

Suppose we could enumerate the numbers between 0 and 1, written as decimals. If this is done, we arrive at a list

$$b_1 = 0.a_{11}a_{12}a_{13}a_{14}\dots$$
$$b_2 = 0.a_{21}a_{22}a_{23}a_{24}\dots$$
$$b_3 = 0.a_{31}a_{32}a_{33}a_{34}\dots$$
$$\vdots$$

The a_{kn} are digits from 0 to 9.—Possibly, we have

$$b_{59284} = 0.5000\dots$$

According to our assumption, the list includes each number between 0 and 1.

© The Author(s), under exclusive license to Springer Nature Switzerland AG 2022
D. D. Spalt, *A Brief History of Analysis*,
https://doi.org/10.1007/978-3-031-00650-0_14

The surprising "argument" that should supply the contradiction to our supposition is as follows:

- Take the number $c = 0.c_1c_2c_3c_4\ldots$, defined by
  c_1 = any number, *but not* a_{11};
  c_2 = any number, *but not* a_{22};
  c_3 = any number, *but not* a_{33};
  etc.
- This number c differs obviously from all b_m. (This c deviates in the m-th digit from b_m, as this digit of b_m is $= a_{mm} \neq c_m$.)
- *Consequently,* the list of the b_m does not contain c.
- But this contradicts the initial *assumption* that we *had* a list b_m of *all* numbers between 0 and 1.
- Contradiction. *Presumed* end of proof.

Every modern book on the mathematical infinite presents this "proof", including the book *Proofs from THE BOOK* from Martin Aigner and Günter M. Ziegler from 1998 as well (pp. 92f). And certainly the authors are excited: star writer David Foster Wallace (1962–2008) e.g. called this proof "both ingenious and beautiful—a total confirmation of art's compresence in pure math" (p. 255). An appraisal such as "ingenious" calls for attention, no doubt about that!

The Impotence of This Reasoning

Let us pause for thought and ask: does this actually convince anyone?
Let us work out the details. At once we realise:

> The above "proof" is not necessarily valid.

(This is no news. I only repeat a well-known reasoning.)—Therefore, I ask two questions.

> (1) *How many decimal numbers with n digits in between* 0 *and* 1 *do exist?*

The answer is straightforward:

> 1. If $n = 1$, we have exactly 10, to wit: 0.0; 0.1; 0.2; ... 0.9.
> 2. If $n = 2$, we have exactly $100 = 10^2$, to wit: 0.00; 0.01; 0.02; ... 0.99. etc.
> n. Consequently, in the general case n, we have 10^n numbers.

And now my second question:

> (2) *How many digits of c are safely determined by this so-called proof?*

The answer to this question is evident: exactly n—one for each digit under consideration. But what is the consequence? Plainly this: the number c defined by this professed proof certainly differs from the first n numbers in the list—but *this does not demonstrate* CONCLUSIVELY that it is not included in the list at all; instead, it *only* proves: $c = b_{m'}$ as well as $m' > n$, nothing else.

To be practical, let $n = 2$. We write down the list of the $10^2 = 100$ numbers in between 0 and 1 with two digits after the decimal point:

$$b_1 = 0.00, \quad b_2 = 0.01 \quad b_{11} = 0.10 \quad \ldots, \quad b_{100} = 0.99.$$
$$b_3 = 0.02 \quad b_{12} = 0.11$$
$$\ldots, \quad \ldots,$$

Let us take $c = 0.12$. Then we know $c \neq b_1$ and $c \neq b_2$; *but* we have $c = b_{13}$. c is not to be found among the first $n = 2$ numbers but later on it *is* included in the list.

This reasoning imputes a (MOST) EXTREMELY STRANGE *hidden assumption* to this "both ingenious and beautiful" "proof". It is this one:

$$10^\infty = \infty.$$

For in fact it is proclaimed: The number c is *not* contained in the list of the b_m. What is true, however, is just this: The number c is *not among the first infinitely many* ("∞") *numbers of the list,* but only subsequently.

However, who is demanding to calculate with infinity in this way? It is true, e.g. **Weierstraß** authorized this—i.e. $10^\infty = \infty$—but at the same time he demanded to exclude infinity ("∞") from analysis (p. 177).

But **Euler** did use infinity in his calculations. He even assumed $i > i - 1$ (p. 77) and moreover

$$10^i > i$$

for infinitely large i.

There is no obligation to think like Euler and also none to think like Weierstraß.

> In mathematics we do not deal with *dogmas* but with *arguments*.

From Chap. 5 we know:

> There exists no conclusive or ultimate reasoning for dealing with the actual infinite in analysis.

The Origin of the "Diagonal Argument"

The "proof", which is described above, was invented by Georg Cantor. Cantor demonstrated thus in 1891 for the third time:

> Theorem. *There exist infinite sets which cannot be related one-to-one to the set of natural numbers.*

Cantor designed his proof slightly differently. (He took only *two* different digits instead of *ten* as we did.—Strikingly, this proof, given for binary numbers instead of decimal numbers, is not only *not conclusive* but also *wrong*—why exactly? And: is it permissible to make a proof about the *real numbers* dependent on *the mode of their representation?*)

The True Understanding of the "Diagonal Argument"

It was Bertrand Russell who unearthed the rational essence of Cantor's proof.

> The *true* theorem of Cantor's proof is this:
> Theorem. *The number of subsets of a set is larger than the number of the elements of this set.*

Let us take an example. The given set may contain two elements, a and b:

$$M = \{ a, b \}.$$

Then M has the following $2^2 = 4$ subsets, first of all the empty set $\{ \} = \emptyset$:

$$\emptyset, \quad \{ a \}, \quad \{ b \}, \quad \{ a, b \}.$$

Hence, there are more subsets than elements: $4 > 2$.

Now we faithfully present Russell's general proof of his theorem (changed, however, are the names of the sets—which Russell calls "classes"). Like Cantor, Russell starts with a *list* of subsets signified by the elements. Russell calls this list a "one-one correlation R".

When a one-one correlation R is established between all the members of M and some of its sub-classes, it may happen that a given member x is correlated with a sub-class of which it is a member; or, again, it may happen that x is correlated with a sub-class of which it is not a member.

We interrupt Russell and illustrate his set-theoretical construction with the help of our example, and that twice.

1. Illustration. The list of the correspondences of the elements and the subsets of $M = \{a, b\}$ might be

$$a \mapsto \{a\}, \qquad b \mapsto \{a, b\}.$$

Then we have for *both* elements a and b that they correspond to a subset that *contains them as an element*.

2. Illustration. A second list might be:

$$a \mapsto \emptyset, \qquad b \mapsto \{b\}.$$

In this case, only element b corresponds to a subset, which contains itself as an element, but element a does not.

Now we continue with Russell's proof.

Let us form the whole class, N say, of those members x which are correlated with sub-classes of which they are not a member.

In our first illustration, we have $N = \emptyset$; in the second, it is $N = \{a\}$. We continue with the proof. Now follows the decisive sentence (I give it *in italics*):

This is a sub-class of M, and it is not correlated with any member of M.
   For taking first the members of N, each of them is (by the definition of N) correlated with some sub-class of which it is not a member, and is therefore not correlated with N.
   Taking next the terms that are not members of N, each of them (by the definition of N) is correlated with some sub-class of which it is a member, and therefore again is not correlated with N.
   Thus no member of M is correlated with N.
   Since R was *any* one-one correlation of all members with some sub-classes, it follows that there is no correlation of all members with *all* sub-classes.

It is extremely impressive how a few words can express a really entangled reasoning. Those who are able to understand this demonstration from Russell may be pleased with their capability of abstract thought.
   If you love formalism, you may write down this proof in the following manner (usually "2^M" denotes the set of all subsets of M; "$M \setminus N$" indicates the difference of the sets M and N):

Let $R : M \longrightarrow 2^M$; therefore $M \ni x \xmapsto{R} Rx \in 2^M$.

Let $N = \{x \mid x \notin Rx\} \subseteq M$. Then we have for all $x \in M$: $Rx \neq N$.

Proof:

1. Let $x \in N$. Then $x \notin Rx$, and consequently, $N \neq Rx$ (for $N \ni x \notin Rx$).

2. Let $x \in M \setminus N$. Then we have $x \notin N$; therefore, $x \in Rx$, and again $Rx \neq N$.

(In both cases, an element of one set is shown,

which is missing in the other set.)

1 & 2: $x \in M \Longrightarrow Rx \neq N$, as was claimed.

In this way, Russell established:

> Theorem. 2^n is always larger than n—even if n is infinite.
> This implies that the infinite cardinal numbers do not have a maximum.

The Significance of the "Diagonal Argument"

That is why the so-called Diagonal Argument from Cantor is a *method of proof* in set-theory. It gives the means of proving e.g. that there always exist larger "cardinalities", even in the infinite.

The "numbers" in set-theory are "cardinal" numbers. In analysis they do not feature (at least not in traditional analysis, in calculus). Calculus needs "numbers for calculations".

Therefore, Russell's proof that $2^n > n$ (which holds *in set-theory*) does not contradict Weierstraß' principle that $10^\infty = \infty$. Quite the contrary: if one accepts Cantor's Diagonal Argument as permissible reasoning *in analysis,* you *must* also accept Weierstraß' dictum that $10^\infty = \infty$, for otherwise the Diagonal Argument fails! This is explained above.

So, if you think analysis (in this sense) and set-theory as one thing, you will be convinced that $10^\infty = \infty$ (in analysis) *and* that $2^n > n$ for all n, including infinite numbers (in set-theory).

Some will like this ... But in some vague sense, this does not look altogether *consistent.* Luckily, one is not *forced* to think in this style.

For the record: on the case of *numbers used for calculating,* i.e. in analysis, you *may* accept the Diagonal Argument. Or you may not. In any case, it is *not conclusive* there. And if you want to relate it to the facts connected to the decimal numbers (*for which it* CLAIMS *validity*), it is even *less conclusive.* For we have seen:

> The Diagonal Argument demands the acceptance of an argument regarding *infinity,* which *is invalid in any finite case.*

As we know, Leibniz once argued in the same way, thereby surprising Johann Bernoulli (p. 55).

It might cross one's mind to interpret this argument as a *proof by induction:*

- Start: choose $c_1 \neq a_{11}$.
- Inductive step: Let $c_n \neq a_{nn}$. Then choose $c_{n+1} \neq a_{n+1,n+1}$.
  —Finished.

An *objection* to this kind of reasoning is obvious: it is *presupposed* that the first part of each number $0.c_1 c_2 \ldots c_n$ formed in this way is contained in the list of the b_m: $0.c_1 c_2 \ldots c_n = b_{m'}$; of course, it is $m' > n$. So what? If *any* first part of c is contained in the list of the b_m—why should this hold no longer for c itself?

If you do not accept the Diagonal Argument in analysis (to repeat: this is as legitimate as is the opposite), you are accepting the validity of

$$10^i > i$$

for infinite numbers used in calculations. Finally, we shall demonstrate where this kind of thinking will lead to.

Paradox I: Conditionally Convergent Series

Let us finish our detour into set-theory and return to analysis. There are certain curiosities that have arisen during the last three and a half centuries that caught the attention of some analysts.

$$1 - \tfrac{1}{2} + \tfrac{1}{3} - \tfrac{1}{4} + \tfrac{1}{5} - + \ldots = \ln 2 .$$

Let us now tackle some of these curiosities in more detail, using the title "paradoxes". The creator of this name will be uncovered later.

In order to prepare for this, we now return to Riemann. We know: Riemann is always one for offering a surprising perspective.

A Mathematical Monstrosity: The Riemann Theorem on Rearrangements of 1854

Riemann proved the following oddity, nowadays called the "Riemann Theorem on Rearrangements", in his habilitation thesis of 1854:

> Theorem. If a convergent series is no longer convergent when its terms are made positive, the terms can be rearranged in such a way that the series converges on any arbitrarily chosen value.

The proof seems to be straightforward. One assumes a *convergent* series of numbers:

$$\sum_{k=1}^{\infty} a_k = s .$$

Infinitely many terms have the sign $+$, and infinitely many have the sign $-$. We call the positive terms of the series b, the negative ones c:

$$\sum_{i=1}^{\infty} b_i = \sum_{\substack{a_k > 0 \text{ and} \\ k=1}}^{k=\infty} a_k \qquad \text{as well as} \qquad \sum_{i=1}^{\infty} c_i = \sum_{\substack{a_k < 0 \text{ and} \\ k=1}}^{k=\infty} a_k .$$

So we have $b_i > 0$ as well as $c_i < 0$. Neither of those series converges:

$$\sum_{i=1}^{\infty} b_i = \sum_{\substack{a_k>0 \text{ and} \\ k=1}}^{k=\infty} a_k = \infty \quad \text{as well as} \quad \sum_{i=1}^{\infty} c_i = \sum_{\substack{a_k<0 \text{ and} \\ k=1}}^{k=\infty} a_k = -\infty.$$

If *both* series had finite sums, then the series $\sum |a_k|$ would also converge, which contradicts our assumption. If only one of these two series converges, the initial series could not converge. That is why none of our partial series converges.

Now Riemann's reasoning: let D, say $D > 0$, be an arbitrarily given value. Then we approach D gradually—at first from below, then from above; and so on, alternating. That is to say, we choose the smallest number n_1 with

$$\sum_{k=1}^{n_1} b_k > D; \quad \text{thereafter, the smallest number } m_1 \text{ with } \quad \sum_{k=1}^{n_1} b_k + \sum_{k=1}^{m_1} c_k < D,$$

etc. The difference to D will never be more than the absolute value of the term that is the one before the last sign change. But $\sum a_k$ converges, and therefore, b_k as well as $|c_k|$ are decreasing with increasing k below any given quantity; and so does the difference to D—which seems to prove all.

This theorem is a monster. Weierstraß obviously did not like it, and he indeed mentions the notion of conditional convergence (see p. 184), but he ignores this concept in his own analysis.

Why is this theorem a monster? Because it assumes that a mathematician is able *to decide infinitely many times* AT WILL to change between the partial series of the b_i and c_i back and forth. "An infinite number of arbitrary choices is an impossibility" says Russell. Infinitely many *actual* decisions have nothing in common with reality.

Note: A proof may be easy, but the proven statement may be absurd.

Mitigation

A theory of infinitely many *single* choices in analysis certainly is an extreme. An essential mitigation is to think about a *regular* change of infinitely many terms of a series. We might have more chance of acceptance if we concentrate on *regular* mathematical constructions—although Weierstraß *might* still have taken a different view.

Let us take an **example.** We start with a series A, halve every term, add both series term-by-term, and regain the initial series, although slightly rearranged. We have:

$$A = \quad \ln 2 = 1 - \tfrac{1}{2} + \tfrac{1}{3} - \tfrac{1}{4} + \tfrac{1}{5} - \tfrac{1}{6} + \tfrac{1}{7} - \tfrac{1}{8} + - \dots$$

Therefore, $\frac{1}{2} \ln 2 = \quad \frac{1}{2} \quad - \frac{1}{4} \quad + \frac{1}{6} \quad - \frac{1}{8} + - \ldots ,$

and consequently $B = \frac{3}{2} \ln 2 = 1 \quad + \frac{1}{3} - \frac{1}{2} + \frac{1}{5} \quad + \frac{1}{7} - \frac{1}{4} + + - \ldots$

$= \quad$ [rearrangement of] $\quad \ln 2 .$

Do we have $A = B$, i.e. $\ln 2 = \frac{3}{2} \ln 2$?

According to Weierstraß, this conclusion is a "fallacy" (p. 177). Why?—Is the Riemann Theorem on Rearrangements an answer?—Is this answer satisfactory?

Paradox II: Methods of Summation

For divergent series like

$$A = 1 - 1 + 1 - 1 + 1 - 1 + - \ldots$$

sometimes other summation methods are in use, e.g. the so-called method of C-1 summation:

$$S^{C1} = \lim_{n \to \infty} \frac{s_1 + s_2 + s_3 + \ldots + s_n}{n} .$$

In case of series A, the partial sums are $s_1 = 1$; $s_2 = 1 - 1$; $s_3 = 1 - 1 + 1$; $s_4 = 1 - 1 + 1 - 1$ etc., that is:

$$s_1 = s_3 = s_5 = \ldots = 1 ,$$

$$s_2 = s_4 = s_6 = \ldots = 0 ,$$

and consequently, the terms of $S^{C\,1}$ are

$$p_1 = \frac{s_1}{1} = 1 ,$$

$$p_2 = \frac{s_1 + s_2}{2} = \frac{1}{2} ,$$

$$p_3 = \frac{s_1 + s_2 + s_3}{3} = \frac{2}{3} ,$$

$$p_4 = \frac{s_1 + s_2 + s_3 + s_4}{4} = \frac{2}{4} = \frac{1}{2} ,$$

$$p_5 = \frac{s_1 + s_2 + s_3 + s_4 + s_5}{5} = \frac{3}{5} ,$$

$$\ldots$$

i.e. $$p_{2n-1} = \frac{n}{2n-1} ,$$

$$p_{2n} = \frac{n}{2n} = \frac{1}{2} ,$$

all together: $$S^{C\,1} = \lim_{n \to \infty} p_n = \frac{1}{2} .$$

With this method, we do get:

$$A = 1 - 1 + 1 - 1 + 1 - 1 + - \ldots = \tfrac{1}{2} \ !?$$

Paradox III: The Convergence of Function Series

Let us take the geometric series

$$s(x) = 1 - x + x^2 - x^3 + x^4 - + \ldots$$

The well-known way to calculate the sum $s(x)$ is to add $s_n(x)$ and $x \cdot s_n(x)$:

$$s_n(x) = 1 - x + x^2 - x^3 + x^4 - x^5 + - \ldots + (-1)^{n-1}x^{n-1}$$

$$\underline{x \cdot s_n(x) = \qquad x - x^2 + x^3 - x^4 + x^5 - + \ldots + (-1)^{n-2}x^{n-1} + (-1)^{n-1}x^n}$$

$$(1+x) \cdot s_n(x) = 1 + (-1)^{n-1}x^n$$

and $\boxed{s_n(x) = \dfrac{1+(-1)^{n-1}x^n}{1+x} \qquad \text{if} \quad x \neq -1.}$

The consequence is

If $\quad |x| < 1 \quad$ we have $\quad s(x) = \lim\limits_{n \to \infty} s_n(x) = \tfrac{1}{1+x}$,

if $\quad x = 1 \quad$ we have $\quad s(1) = \tfrac{1}{2}\,(1 - 1 + 1 - 1 + - \ldots)$, which does not exist.

Yet, "does not exist" is not a beautiful result of a calculation.

Paradox IV: The Term-by-Term Integration of Series

Let us take the series of functions

$$f_n(x) = \frac{nx}{1 + n^2 x^4} \qquad \text{with} \quad 0 \le x \le 1.$$

As $\lim\limits_{n \to \infty} \dfrac{nx}{1+n^2x^4} = 0$ if $0 \le x \le 1$, we usually conclude for the limit $n \to \infty$:

$$f(x) = \lim_{n \to \infty} f_n(x) = \lim_{n \to \infty} \frac{nx}{1 + n^2 x^4} = 0. \tag{$\ddagger\ddagger$}$$

But if $x_n = \frac{1}{\sqrt{n}}$ (and consequently $\lim\limits_{n\to\infty} x_n = \lim\limits_{n\to\infty} \frac{1}{\sqrt{n}} = 0$), the value of $f(x)$ at the value $0 = \lim\limits_{n\to\infty} \frac{1}{\sqrt{n}}$ is

$$\lim_{n\to\infty} f_n\left(\frac{1}{\sqrt{n}}\right) = \lim_{n\to\infty} \frac{\sqrt{n}}{1+\frac{n^2}{n^2}} = \lim_{n\to\infty} \frac{\sqrt{n}}{2} = \infty\ !$$

We remember: in these cases Cauchy concludes ∞ to be a value of the function $f(x)$ at the value $x = 0$ (p. 124). However, present-day analysis handles things differently and demands:

$$f(0) = \lim_{n\to\infty} f_n(0) = \lim_{n\to\infty} \frac{n\cdot 0}{1+n^2\cdot 0} = \frac{0}{1} = 0,$$

unambiguous and crystal-clear.

If we have, following ‡‡, $f(x) = 0$ for all $0 \leq x \leq 1$, integration is easy:

$$\int\limits_0^1 f(x)\, dx = \int\limits_0^1 0\cdot dx = 0.$$

But let us integrate term-by-term:

$$\int\limits_0^1 f_n(x)\, dx = \int\limits_0^1 \frac{nx}{1+n^2x^4}\, dx\ .$$

Substituting $z = nx^2$ gives $dz = 2nx\, dx$ and we arrive at

$$= \tfrac{1}{2}\cdot \int\limits_0^n \frac{dz}{1+z^2} = \tfrac{1}{2}\cdot \arctan z\ \Big|_{z=0}^{z=n} = \tfrac{1}{2}\cdot \arctan n\ .$$

With the help of $f(x) = \lim f_n(x)$, we conclude

$$\int_0^1 f(x)\, dx = \lim_{n\to\infty} \int_0^1 f_n(x)\, dx = \lim_{n\to\infty} \tfrac{1}{2}\cdot \arctan n = \tfrac{\pi}{4}\ .$$

This contradicts the previous result. Consequently, the first equality in the last line is wrong—the others being indubitable. So, for this series, it is *not* possible to interchange integration and limit.

Conclusion: We need a *theorem* to tell us, *in which cases* the interchange of integration and limit is allowed. Naturally, it would be more attractive if *the calculation itself* would make the problem *visible*. Some analysts enjoy calculating more than hunting for theorems that allow them to calculate.

Is an Analysis Without Those Paradoxes Possible?

A Source from the Years 1948–53

Curt Schmieden (1905–91) was an eminent mathematician. In 1934, he became a professor at the Technical University of Darmstadt, where he was awarded a chair in mathematics in 1937, and in 1957/58 he was the rector of this university. In a manuscript dating from 1948–53, Schmieden commented on examples of this kind (ALL *emphases* added):

> Such examples could be given endlessly; the deeper one penetrates into mathematics, the more such paradoxes emerge—even if one completely disregards set-theory.
>
> The most thrilling aspect of this kind of mathematics is that in spite of those paradoxes one always arrives at a "true" result for each concrete problem. Consequently, it does not surprise that some mathematicians decide at some point that *the naive intuition about infinity does no longer suffice in some way or other,* whereas working along definite rules in regard to infinity makes it possible to tame those quantities.
>
> *Nevertheless, there remains a feeling of discomfort. Also the question remains,* WHETHER IT IS NOT POSSIBLE AFTER ALL TO CONSTRUCT ANALYSIS IN SUCH A WAY THAT IT WORKS ALONGSIDE FIXED RULES AS IT HAS BEEN DOING SO FAR, AND THIS IN SUCH A WAY THAT *the kinds of aforementioned* PARADOXES DO NOT EMERGE. *Thus our naive intuition* which according to Goethe, more often than not wishes "to represent the infinite with the help of the finite" [1], *would be granted.*

Schmieden, being "fully aware of the imperfections of this attempt" stated some foundational principles along those lines. Among them was the following that is obviously inspired by Johann Bernoulli and Euler but which also surpasses both conceptually:

> Reason forcibly demands that there *must* exist infinitely large numbers, which can *only* be meaningfully defined as follows:

> An infinitely large natural number is a natural number, which *cannot* be grasped by an unlimited continuation of the process of counting, for counting is necessarily bound to the degree of finiteness.

> Consequently, to arrive at such a "number", it requires, figuratively speaking, a jump, or as a metaphor: infinitely large numbers form the "horizon of finiteness".

The idea is:

> If we conceive an infinitely large number called Ω (capital omega), reached by such a jump, obviously nothing speaks against but everything for it that we are allowed to calculate with such a "number" in just the same way as with a usual number.

Basically, Johann Bernoulli and Euler had shared this opinion. A little later Schmieden gives the following definition:

> $\omega = \frac{1}{\Omega}$ (small omega) is an infinitely small number which is smaller than each positive rational number and which is defined by the level of finite numbers. (ω is a "zero number".)

---

[1] This is the exact translation of Goethe's words.

In the same way as for any infinitely large number, there exists one of a larger *order*[2], and there exists for any zero number one of a smaller *order*[2].

Therefore, our ordinary zero known from conventional analysis is only allowed to be used as another notation of the identity $a \equiv a$.

The latter would have pleased Frege (p. 195), and Weierstraß, too, would have given his approval in regard to such precision in relation to equality.

Finally, there is Schmieden's foundational equation:

> We define
>
> $$\lim_{n \to \infty} n = \Omega \ ;$$
>
> each other limit has to be related to this expression.

(The sign "∞" is not a good choice here and should have been avoided altogether.)

It is fairly evident that it is possible to construct from infinitely large natural numbers suitable rational numbers, e.g. $\frac{1}{\Omega}$ or $1 + \frac{1}{\Omega}$ etc. The essential question is: *Is this enough for analysis?* After all, we now have "infinitely near" numbers, and this also in the neighbourhood of each rational number. In Schmieden's words:

> Around each ordinary rational number there exists an ω-sphere which does not include any other ordinary rational number, but which already contains infinitely many rational numbers of the second class [i.e. level] in the lowest ω-level.
>
> With the help of the enlarged notion of rational numbers, which inevitably follows from the introduction of Ω, we reach the seemingly paradoxical result that analysis, if founded on Ω-numbers, gets along with these [Ω-]rational numbers, and even more—that no other numbers can occur, because each number defined via the usual limit arises in our system as an Ω-rational number.

Here we have the essence of Schmieden's idea: instead of the real numbers (following Müller or Bertrand or Cantor or Dedekind), we take "Ω-rational" numbers, e.g. $q \pm \frac{1}{\Omega}$ or $q \pm \frac{5\Omega^2 - 7\Omega + 3}{\Omega^3 - \frac{3}{4}}$, etc. These can be used in calculations in the usual manner. The "Ω-rational" numbers should be enough because—as we know since Euler—analysis consists essentially of calculations, especially of calculations with the infinite.

Schmieden Dissolves the Paradoxes

It was the aim of Schmieden to dissolve such "paradoxes" as shown above with the help of more precise calculations—i.e. with his Ω-numbers. Schmieden states:

> We especially inspect those instances where analysis puts up signs with the warning: here it is forbidden to calculate as in the finite case!

In what follows, representative examples of his method are shown.

---

[2] emphases added

Resolving paradox I   (p. 227).

Take the series

$$\ln 2 = 1 - \tfrac{1}{2} + \tfrac{1}{3} - \tfrac{1}{4} + \tfrac{1}{5} - \tfrac{1}{6} + \tfrac{1}{7} - \tfrac{1}{8} + - \ldots$$

Schmieden now takes care in regard to the infinite ("$+ - \ldots$") and calculates precisely:

$$\ln 2 = \sum_{n=1}^{\Omega} \tfrac{(-1)^{n-1}}{n} = 1 - \tfrac{1}{2} + \tfrac{1}{3} - \tfrac{1}{4} + \tfrac{1}{5} - \tfrac{1}{6} + \tfrac{1}{7} - \tfrac{1}{8} + - \ldots + \tfrac{1}{\Omega-1} - \tfrac{1}{\Omega} =: A .$$

Here Ω is assumed to be even. But careful: "$+ - \ldots$" has *different* meanings in the last two lines of formulae. (a) In the first line, "$+ - \ldots$" indicates "and so on" without last term. (b) In the second line, "$+ - \ldots$" marks a gap, like in $1 + \tfrac{1}{2} + \tfrac{1}{4} + \ldots + \tfrac{1}{2^n}$.

Now Schmieden rearranges the series A and supposes that Ω divided by 4 leaves no remainder. (Why not? If it is useful! At a pinch, make $\Omega' = 4\Omega$.) He then realizes the sum formula of the rearranged series:

$$\sum_{n=1}^{\Omega/4} \left(\tfrac{1}{4n-3} + \tfrac{1}{4n-1} - \tfrac{2}{4n} \right) = 1 + \tfrac{1}{3} - \tfrac{1}{2} + \tfrac{1}{5} + \tfrac{1}{7} - \tfrac{1}{4} + + - \ldots + \tfrac{1}{\Omega-3} + \tfrac{1}{\Omega-1} - \tfrac{1}{\Omega/2} =: B .$$

We *see:*

It is true, A and B have the same positive terms (all odd denominators, including $\Omega - 1$), but the second half of the negative terms of A (even denominators) are missing in B, up to Ω:

$$-\frac{1}{\Omega/2+2} - \frac{1}{\Omega/2+4} - \frac{1}{\Omega/2+6} - \ldots - \frac{1}{\Omega} = - \sum_{n=\frac{\Omega}{4}+1}^{\Omega/2} \tfrac{1}{2n} =: -C ,$$

and we have

$$A = B - C .$$

There you are! *That is why* we have $B > A$!

> B is not at all a rearrangement of A, for it has infinitely *fewer* (negative) terms.

Schmieden even calculates the value of this difference C. This he does by integration—using the fact that in case of a continuous function and an infinitely small interval the estimated value of the area under this function only differs from the integral by an infinitely small amount. We denote this just like we did in the case of Johann Bernoulli (p. 62) with $\approx$ and observe that $\omega = \frac{1}{\Omega}$ is infinitely small. We are going to calculate

$$C = \sum_{n=\frac{\Omega}{4}+1}^{\Omega/2} \frac{1}{2n} \cdot \frac{\omega}{\omega}.$$

Schmieden defines $z = 2n\omega$, consequently $dz = 2\omega$ (as n always grows by 1), and therefore $\frac{\omega}{2n\omega} = \frac{1}{2}\frac{dz}{z}$. But $\frac{1}{2n\omega}$ is continuous (in the sense of ε-δ). Lower limit: $z = 2n\omega = 2\left(\frac{\Omega}{4}+1\right) \cdot \omega = \frac{1}{2} + 2\omega$, upper limit: $z = 2n\omega = 2\left(\frac{\Omega}{2}\right) \cdot \omega = 1$, and therefore:

$$C \approx \frac{1}{2} \cdot \int_{z=\frac{1}{2}+2\omega}^{1} \frac{dz}{z} \approx \frac{1}{2} \cdot (\ln 1 - \ln \tfrac{1}{2}) = \tfrac{1}{2}(\ln 1 - \ln 1 + \ln 2) = \tfrac{1}{2}\ln 2.$$

So we get in fact

$$B - C \approx \tfrac{3}{2}\ln 2 - \tfrac{1}{2}\ln 2 = \ln 2 = A$$

$$resp. \qquad A - B = -C \approx -\tfrac{1}{2}\ln 2,$$

as we expected. Nothing of the kind $\frac{3}{2}\ln 2 = \ln 2$!
  We record:

> The investigation of *regular* "rearrangements" of conditionally convergent series leads to comparable results: *seemingly* paradoxical equations of conventional analysis are explained in calculations with Ω-numbers through the omission of infinitely many terms with a finite sum.

Besides, the concept of "conditional" (i.e. not "absolute") convergence is dispensable. Instead, all series are treated equally, like finite sums. (This would have pleased Weierstraß.)

Interestingly, in their publication from 1958, Schmieden and Laugwitz hint at the fact that "nevertheless, the difference between absolute and conditional convergence remains important for technical calculations, insofar as absolutely convergent series do allow arbitrary rearrangements without further considerations". Obviously, this sentence is a concession by the authors and is probably due to the pressure exerted by the editors of the journal—because essentially it contradicts the foundational view of the authors: in their way of *precisely* calculating, all series are treated equally.

Resolving paradox II (p. 223).

Schmieden considers the C-1 Summation of the series

$$A = 1 - 1 + 1 - 1 + 1 - 1 + - \dots$$

from p. 223. There we examined the expression

$$S^{C\,1} = \lim_{n \to \infty} \frac{s_1 + s_2 + s_3 + \ldots + s_n}{n}.$$

Schmieden presents this series more conveniently via Ω instead of lim and ∞:

$$A^{C\,1} = \frac{s_1 + s_2 + s_3 + \ldots + s_\Omega}{\Omega}.$$

This fraction he writes down in more detail. Thereby he pays attention to the following: (i) The first 1 of series A is included in *all* partial sums s_k; this amounts to the first term being $\frac{\Omega}{\Omega}$. (ii) The second 1 of series A is only included in $\Omega - 1$ of the partial sums s_k (i.e. not in s_1); this gives the second term as $\frac{\Omega-1}{\Omega}$; etc.

$$A^{C\,1} = \frac{\Omega}{\Omega} - \frac{\Omega-1}{\Omega} + \frac{\Omega-2}{\Omega} - + \ldots + (-1)^\Omega \frac{\Omega-(\Omega-2)}{\Omega} + \frac{(-1)^{\Omega+1}}{\Omega}$$

$$= 1 - (1 - \omega) + (1 - 2\omega) - + \ldots + (-1)^\Omega \cdot 2\omega \quad + (-1)^{\Omega+1} \cdot 1\omega$$

He then considers the difference of $A - A^{C\,1}$ step-by-step, going from term to term, as follows:

$$A - A^{C\,1} = \underbrace{0}_{0} \; -\omega +2\omega -3\omega +4\omega \quad \ldots \quad +(-1)^{\Omega+1}(-\omega)$$

$$\underbrace{-\omega}$$
$$\underbrace{\omega}$$
$$\underbrace{-2\omega}$$
$$\underbrace{2\omega}$$
$$\ldots$$

$$= \begin{cases} -\frac{\Omega}{2}\omega = -\frac{1}{2} & \text{if} \quad \Omega \quad \text{even, then} \quad A = 0 \\ +\frac{\Omega-1}{2}\omega \approx +\frac{1}{2} & \text{if} \quad \Omega \quad \text{odd, then} \quad A = 1. \end{cases}$$

All this is very accurate. *Note the sum A has the value* $\frac{1}{2}$, *but we have* $A^{C\,1} = -\frac{1}{2}$ *if Ω is even, if not we have* $A^{C\,1} \approx \frac{1}{2}$. *The sum A* ALWAYS *has one of the values* 1 *or* 0.—No paradox whatsoever.

The following question presents itself:

> *What constitutes the sum $A^{C\,1}$ really?*

Schmieden just calculates. For the general

$$A = a_0 + a_1 + a_2 + \ldots + a_{\Omega-1}$$

he takes, as we have already seen,

$$A^{C\,1} = \tfrac{1}{\Omega} \sum_{n=0}^{\Omega-1} s_n = \tfrac{1}{\Omega} \left(\Omega a_0 + (\Omega - 1)a_1 + (\Omega - 2)a_2 + \ldots + (\Omega - (\Omega - 1))a_{\Omega-1} \right) .$$

Especially, if

$$A(x) = 1 + x + x^2 + \ldots + x^{\Omega-1},$$

he substitutes $a_n = x^n$ and gets

$$A^{C\,1}(x) = \tfrac{1}{\Omega} \sum_{n=0}^{\Omega-1} s_n = \tfrac{1}{\Omega} \left(\Omega \cdot 1 + (\Omega - 1)x + (\Omega - 2)x^2 + \ldots + (\Omega - (\Omega - 1))x^{\Omega-1} \right)$$

$$= 1 - x(1 - \omega) + x^2(1 - 2\omega) - + \ldots + (-1)^{\Omega-1}x^{\Omega-1}(1 - (1 - \Omega)\omega)$$

$$= \sum_{n=0}^{\Omega-1}(-1)^n x^n \quad + \quad \omega \sum_{n=0}^{\Omega-1}(-1)^{n+1}n x^n$$

$$= \quad s_\Omega(x) \quad + \quad \omega \sum_{n=0}^{\Omega-1}(-1)^{n+1}n x^n .$$

Therefore, we have

$$s_\Omega(x) - A^{C\,1}(x) = \omega \sum_{n=0}^{\Omega-1}(-1)^n n x^n = K .$$

K generates the convergence of $s_\Omega(x)$ to $A^{C\,1}(x)$, but only in the infinite: As the factor ω in K shows, a finite number of terms of K will have sum ≈ 0.

Resolving paradox III   (p. 224).

From the formula

$$s_n(x) = \frac{1 + (-1)^{n-1}x^n}{1 + x} \qquad \text{if} \qquad x \neq -1,$$

Schmieden directly concludes

$$s_\Omega(x) = \tfrac{1}{1+x} \left(1 + (-1)^{\Omega-1}x^\Omega \right) . \tag{§§}$$

1. If $|x| < 1$ and $|x| \not\approx 1$, we get $x^\Omega \approx 0$, and consequently,

$$s_\Omega(x) \approx \tfrac{1}{1+x} .$$

2. If $x = 1$, we get for an even or odd number Ω of terms:

$$s_\Omega(1) = \tfrac{1}{2}\left(1 + (-1)^{\Omega-1}\right) = \begin{cases} 1 & \text{if } \Omega \text{ is odd} \\ 0 & \text{if } \Omega \text{ is even.} \end{cases}$$

3. If $x < 1$ and $x \approx 1$, the tricky x^Ω is calculated by means of a trick. Like Euler, Schmieden takes for $v > 0$:

$$x := 1 - \xi \omega^v$$

where ξ is finite and positive. He gets (remember from p. 77: instead of $\lim\limits_{n\to\infty} \left(1 + \tfrac{x}{n}\right)^n = e^x$, Schmieden has of course $(1 + \omega x)^\Omega \approx e^x$) with the help of the following subsidiary calculation:

$$x^\Omega = \left(1 - \xi\omega^v\right)^\Omega = \left(1 - \omega \cdot \xi\omega^{v-1}\right)^\Omega \approx e^{-\xi\omega^{v-1}} \begin{cases} \approx 1 & \text{if } v > 1, \\ = e^{-\xi} & \text{if } v = 1, \\ \approx 0 & \text{if } v < 1, \end{cases}$$

and finally for §§ (as $1 + x \approx 2$), the result

$$s_\Omega(x) \approx \begin{cases} \tfrac{1}{2}\left(1 + (-1)^{\Omega-1}\right) & \text{if } v > 1, \\ \tfrac{1}{2}\left(1 + (-1)^{\Omega-1}e^{-\xi}\right) & \text{if } v = 1, \\ \tfrac{1}{2} & \text{if } 0 < v < 1. \end{cases}$$

The case $|x| > 1$ is judged by Schmieden as being "nonsensical" (Fig. 14.1).

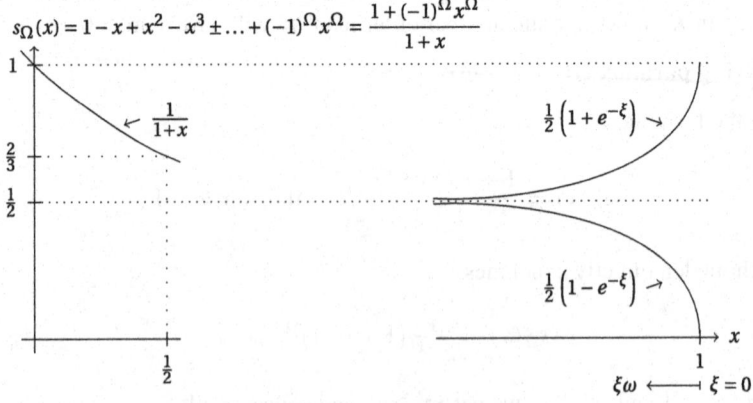

$$s_\Omega(x) = 1 - x + x^2 - x^3 \pm \ldots + (-1)^\Omega x^\Omega = \frac{1 + (-1)^\Omega x^\Omega}{1 + x}$$

Fig. 14.1 The function $s_\Omega(x) = 1 - x + x^2 - + \ldots$ in $[0, 1]$, at right for $x \overset{<}{\approx} 1$

Everything is exactly explained: if $\nu = 1$ and:

(i) Ω is odd, the function $s_\Omega(x)$ increases from the value $\frac{1}{2}$ near $x = 1$ as e-function $\frac{1}{2}\left(1 + e^{-\xi}\right)$ up to the value 1 for $x = 1$ (i.e. up to $\xi = 0$).

(ii) Ω is even, the function decreases from the value $\frac{1}{2}$ near $x = 1$ as e-function $\frac{1}{2}\left(1 - e^{-\xi}\right)$ up to the value 0 for $x = 1$.

In other words: depending on whether Ω is odd or even, the value of the function *near $x = 1$ decreases from $\frac{1}{2}$ to 0 or increases from $\frac{1}{2}$ to 1. This is by no means indeterminate!* Of course, some calculation is needed—in mathematics, you do not get anything without work. The distinction between Ω being odd or even shows the different *possibilities* of the results of your calculation.

Resolving paradox IV   (p. 224).

Only Eq. ‡‡ is thorny. Schmieden considers yet again the details and studies

$$f_\Omega(x) = \frac{\Omega x}{1 + \Omega^2 x^4}.$$

As before, he takes $x = \xi\omega^\nu$ and gets

$$f_\Omega\left(\xi\omega^\nu\right) = \frac{\xi\omega^{\nu-1}}{1 + \xi^4\omega^{4\nu-2}} = \frac{\xi\Omega^{1-\nu}}{1 + \xi^4\Omega^{2-4\nu}} \approx \begin{cases} 0 & \text{if } \nu > 1, \\ \xi & \text{if } \nu = 1, \\ \xi^{-3}\Omega^\alpha, (0 < \alpha < 2), & \text{if } \frac{1}{3} < \nu < 1, \\ \quad \text{as } \frac{\xi\Omega^{1-\nu}}{1+\xi^4\Omega^{2-4\nu}} \sim \Omega^{1-\nu-2+4\nu} \\ \quad\quad\quad\quad\quad\quad\quad = \Omega^{3\nu-1} \\ \xi^{-3} & \text{if } \nu = \frac{1}{3}, \\ 0 & \text{if } 0 < \nu < \frac{1}{3}. \end{cases}$$

Therefore near 0 in the case $\frac{1}{3} \leqq \nu \leqq 1$, we unambiguously obtain

$$f_\Omega \neq f \ !$$

Schmieden also deals with delta-functions (indeed, if you know Ω-numbers, you have *true* delta-*functions!*). This would lead us astray, but one example of an elementary delta-function will nevertheless be shown below.

The First Formal Version of a Nonstandard-Analysis in the Year 1958

In 1954, the then young Detlef Laugwitz (1932–2000) was handed a manuscript by Carl Friedrich von Weizsäcker (1912–2007) to report on it in his seminar in

Göttingen. After Laugwitz' affirmative reaction to it, Weizsäcker brought him into contact with the author Curt Schmieden. This was the beginning of a very fruitful cooperation, and in 1962, at the age of 29, Laugwitz took up a position as chair in mathematics at the University of Darmstadt.

Four years earlier, in 1958, and after some quarrels behind the scenes, the renowned journal *Mathematische Zeitschrift* had published an article by Schmieden and Laugwitz in its volume 69. Later it turned out to be the first publication in a field nowadays called "nonstandard-analysis".

In it, the two authors expressively emphasize that:

> It should be noticed that not only a rebuilding or a mere modification of the conventional analysis emerges here, but that a true enlargement is also produced.

Right at the start, they showed that their new analysis incorporates the species of Dirac's delta-functions as true "functions", without the necessity of modifying the notion of a function. As an example of a delta-function, they presented

$$\delta(x) = \frac{1}{\pi} \cdot \frac{\Omega}{1 + \Omega^2 x^2} \, .$$

The typical features of a delta-function are easily confirmed: $\delta(x) \approx 0$, except when $x \approx 0$—where $\delta(x)$ is infinitely large, and $\int\limits_{-\infty}^{+\infty} \delta(x)\,dx = 1$.

The Foundation in the Year 1958

Laugwitz formulated some of Schmieden's principal ideas in the language of the then fashionable algebra. He referred to the definitions of Cantor. Cantor had defined *convergent sequences of rational numbers* as novel numbers, and he had named these sequences "limits" (p. 191). The duo Schmieden/Laugwitz proposed to take *the totality* of sequences of rational numbers and to define EACH *of these sequences as an individual "number"*—independent of their convergence, just: *all* these sequences! As a name for the sequences, they chose "Limit" (as Cantor did, but with a capital "L" as a distinction) as well as "Ω-number".

$$(a_n)_n = a_\Omega = \mathop{\mathrm{Lim}}_{n=\Omega} a_n \, ,$$

especially $(n)_n = \Omega$, $\quad (\frac{1}{n})_n = \omega$ and also $((-1)^n)_n = (-1)^\Omega$.

That is to say, Schmieden's "jump" from the process of counting 1, 2, 3, ... to Ω (see p. 226) now obtains as its final destination the *complete sequence* $\Omega = (n)_n$! Even this divergent sequence defines an Ω-number, of course an infinitely large one.

The preceding number to Ω, i.e. $\Omega - 1$, is the sequence $(n-1)_n = (0, 1, 2, 3, \ldots)$; its successor, i.e. $\Omega + 1$, is the sequence $(2, 3, 4, \ldots)$, etc.

The discombobulating facts in favour of conventional analysis (that the new numbers comprise zero divisors and therefore do not constitute a field—and consequently division is not always possible—as well as the missing of their "total order") were clearly articulated in this chapter, but not a single word of valuation was added.

For example, we have

$$(0, 1, 0, 1, 0, 1, \ldots) \cdot (1, 0, 1, 0, 1, 0, \ldots) = 0,$$

although neither factor being $0 = (0, 0, 0, \ldots)$. And, moreover, we are not able to call one of these two factors greater than the other—although obviously they are different.

Further Peculiarities of the New Analysis in the Year 1958

The authors also clearly identified those facts that appear in this new "enlargement" of analysis *differently* from in conventional analysis:

- There are three different equivalence relations as orders of magnitude: 1. "finitely equal", 2. "having the same magnitude" and 3. "having the same order of magnitude".

- **Some theorems are stated and proved, which are *wrong* in conventional analysis but *true* in the new enlargement.**
  1. *Every limit exists.* This is just the new definition of "number".
     In detail (in the case of sequences of Ω-numbers, we write their "index" *above*): The limit of a sequence of Ω-numbers $(a_\Omega^{(p)})_p$ is defined for each "positive integer" as well as for each "infinitely large" Ω-number g_Ω, i.e. $g_\Omega = (g_n)_n$ where $g_n > g_k > 0$ if $n > k$. The limit

     $$\operatorname*{Lim}_{p=g_\Omega} a_\Omega^{(p)} = a_\Omega^{(g_\Omega)} = b_\Omega$$

     is defined as the sequence of components

     $$b_n = a_n^{(g_n)},$$

     i.e. by the "diagonal sequence" defined from the sequence $(a_\Omega^{(p)})_p$ of Ω-numbers $a_\Omega^{(p)}$.—Examples: (a) In case of the constant sequence $a_\Omega = \left(\frac{1}{n}\right)_n = \omega$, we have $\lim_{p=\Omega} a_\Omega^{(p)} = \left(\frac{1}{n}\right)_n = a_\Omega = \omega$, as it should be; and $\lim_{p=2\Omega} a_\Omega^{(p)} = \left(\frac{1}{2n}\right)_n = \frac{1}{2}\omega$. (b) The sequence $b_\Omega^{(p)}$ with the pth term

$b_\Omega^{(p)} = \left(\frac{1}{p\cdot n}\right)_n$ is $\lim_{p=\Omega} b_\Omega^{(p)} = \left(\frac{1}{n\cdot n}\right)_n = \frac{1}{\Omega\cdot\Omega} = \omega^2$ and $\lim_{p=2\Omega} b_\Omega^{(p)} = \left(\frac{1}{2n\cdot n}\right)_n = \frac{1}{2\Omega\cdot\Omega} = \frac{1}{2}\omega^2$—right?

2. *Any two limits are interchangeable.* The essential reason for this is that limits are only indicated but not calculated.

   In detail: Given a double sequence $(a_\Omega^{p,q})_{p,q}$ of Ω-numbers $a_\Omega^{g_n,h_n}$, the limit is given by the sequence of the components $a_n^{g_n,h_n}$.

   Example:

$$a^{p,q} = \frac{1}{1 + \frac{p}{q}}.$$

(The given Ω-numbers are constant sequences of rational numbers with the constant components $\frac{1}{1+p/q}$.)
Conventionally, we have

$$\lim_{q\to\infty}\lim_{p\to\infty} a^{p,q} = 0, \qquad \text{but} \qquad \lim_{p\to\infty}\lim_{q\to\infty} a^{p,q} = 1.$$

However, for Ω-*numbers*, independently from the order, we have

$$\operatorname*{Lim}_{\substack{p=g\Omega\\q=h\Omega}} a^{p,q} = \frac{1}{1 + \frac{g\Omega}{h\Omega}}.$$

If you choose $p = g\Omega$ (conventionally: $p \to \infty$) for taking the first limit, you obtain

$$\frac{1}{1+\frac{g\Omega}{q}},$$

i.e. an infinitely small number for each finite $q = 1, 2, 3, \ldots$ Yet, if you then choose for the other limit $h\Omega = g\Omega$, you nevertheless get the finite value

$$\frac{1}{1+\frac{g\Omega}{h\Omega}} = \frac{1}{1+\frac{g\Omega}{g\Omega}} = \frac{1}{1+1} = \frac{1}{2},$$

which says that the infinitely many small numbers $\frac{1}{1+\frac{g\Omega}{q}}$ nevertheless have a finite limit of the value $\frac{1}{2}$.

3. *Divergent series are of equal rank to the others,* as in the new version of analysis the calculations are the same: in the finite and in the infinite.
   As an example, the famous series:

$$\sum_{p=1}^{\Omega}\frac{1}{p(p+1)} = \sum_{p=1}^{\Omega}\left(\frac{1}{p} - \frac{1}{p+1}\right) = \sum_{p=1}^{\Omega}\frac{1}{p} - \sum_{p=2}^{\Omega+1}\frac{1}{p} = 1 - \frac{1}{\Omega+1} \approx 1.$$

A convergent series on the left is split into the difference of two divergent series, and then this difference is calculated to be a finite value. Conventionally, an absolute no-go!

4. *The limit function of a sequence of continuous functions is continuous.* The authors' comment: "This theorem has no simple analogy in conventional analysis". There is no reference made to the "Cauchy Sum Theorem" (p. 138). As an example, the sequence of functions x^n, for the closed interval $0 \leq x \leq 1$, is given (already treated above on p. 140).

- The *Weierstraß Approximation Theorem* was also obtained for the new version of analysis.

- *Differentiation* and *integration* were briefly mentioned. However, a "differentiable" function does *not* need to have a derivative *for each value!* ("The 'derivative' might be 'irrational'".)

- The *Mean Value Theorem* is proved (for "normal" functions).

Naturally, the notion of "function" is precarious. *Of course,* complete freedom in the sense of Bolzano (p. 111) could not be permitted. After all the starting point of Schmieden/Laugwitz is conventional analysis (which they call "usual")—and not a completely new, extravagant shape of analysis. That is why they restrict themselves to such functions that are "already completely defined, if their values are given for the rational numbers of the domain". Sensibly, the description of the values of the function has to be given in a standardized mathematical language.

Finale

Curt Schmieden was a practitioner of calculating. As a consequence, he had the idea of practicing analysis *by calculating:* no complicated general theorems that state the rules of calculation or rule out certain techniques of calculation—just *calculate* and, besides, treat the infinite (i.e. the *specifically* analytical) in just the same way as the finite. (He expressed it this way: "*calculating* analysis".) The result of the calculation has to be assessed according to the requirements of the problem at the end of that calculation.

That this is *permissible* was shown to him by his calculations ("exactly as in the finite"—what should be wrong with this?) as well as by his results. To anchor his technique of "Omega Analysis"—an *enlarged* version of "Value Analysis"—reliably within the current foundational principles of mathematics, he left to others.

His enthusiastic colleague and co-worker Detlef Laugwitz gained credit for his precise algebraic foundation of Schmieden's ideas. This was first published in the most general version in an article from 1958, which was signed by both of them.

In 1978, Laugwitz published a stronger algebraic construction, developed from Schmieden's essential ideas, and in 1986, he presented a new version, which incorporates some features of formal logic.

Laugwitz was always aware that Schmieden's Ω-numbers are not a field because one cannot divide by all non-zero numbers. This seemed to be a problem, especially at a time, when all mathematics was dominated by Nicolas Bourbaki's idea of structure. Subsequently, in 1978, Laugwitz constructed, following others, a field $^*K$ by algebraic means. Later, in 1986, Laugwitz *pretended to, but did not clearly define* a set $^\Omega K$ with Ω-numbers. Instead of defining this, he changed to a formal language and developed a theory where "the rules of an ordered field" are valid.

These attempts to represent Schmieden's ideas within the actual methodology of calculus could not do justice to Schmieden's fundamental concerns. In a *field* that includes Ω-numbers, it *must* be settled, whether $\Omega/2$ is a natural number or not. The mathematician might lack the means to *decide* which it is—but it will definitely *be* one of the two. But this does not fit to Schmieden's essential idea. Schmieden really *left open* both possibilities—and if it were necessary, he distinguished both cases. Methodical stipulations of this kind—even more if they were bound to be unknown in every detail—were surely not his aim.

Foundational Problems

This first publication of a nonstandard analysis by these two men was soon followed by versions from other authors.

The first was by Abraham Robinson (1918–74), a model theorist who also coined the name "nonstandard-analysis". This approach to nonstandard analysis requires the knowledge of logics as a precondition to deal with continuous functions.

Willem A. J. Luxemburg (1929–2018) confined himself to algebraic constructions, essentially relying on ultra-filters for his definition of a field with "infinite" numbers.

In 1977, Edward Nelson (1932–2014) published a first axiom system for nonstandard analysis. Other versions followed.

This means that another mathematical theory (logics, universal algebra) is always needed to provide nonstandard analysis with suitable "numbers". This is not satisfactory and is also a severe obstacle for an easy acceptance of this novel approach to analysis, unless you simply confine yourself to an axiom system.

Axiomatics

When inspecting the actual textbooks of analysis, it appears that there is no adequate consideration paid to the foundations. Generally, the real numbers are not introduced constructively, as was taught by Müller, Bertrand and Dedekind or by Cantor and Heine, but instead, following Hilbert, axiomatically. This helps to save time.

Well, possibly, this fits excellently to the new role of mathematics as a security police for all sciences, which was promulgated by Hilbert in 1917 (p. 209) in order to teach the unconditional (at least provisional) acceptance of *arbitrary* axiom systems detached from any informal substantial justifications.

If it is so, there is then no argument to be put forward against the idea of taking an axiom system for nonstandard numbers (today usually called "hyper-real numbers")—possibly apart from the fact that the elaboration of a nonstandard analysis *differs* in some respects from standard analysis. (This has been shown with the help of some "paradoxes" as well as generalized: see from p. 235.) This challenges one's independent thought as well as impedes the usage of textbooks: you always have to check *which* version of analysis the chosen author likes. Standardization of the curriculum facilitates lecturing, especially if many students have to be educated. In small circles of specialists, subtle discussions will be eased.

A Path to Independency

Schmieden's intention had been to rescue conventional analysis from its "para-doxes" as well as to give it a better form: fewer theorems, more computations.

An opposite development started in the 1980s. It continued for some decades in small circles. Under the (for the laymen astonishing) title "constructive nonstandard-analysis", the theory was developed further as a special field in its own right in which more advanced methods (like sheafs and topoi) were implemented. This leads to a departure from the first origins of nonstandard analysis as well as from classical calculus, but it undoubtedly created some marvellous mathematics.

After the discovery of Weierstraß' construction of the real numbers in 2016, it may well be possible to develop a new kind of approach to nonstandard analysis.

Nonstandard-Analysis and the History of Analysis

Not Schmieden and Laugwitz, but Robinson was the one who immediately started to relate the new theory to historical texts. The theorem nowadays called the "Cauchy Sum Theorem" (p. 138) came up fairly early for discussion. In his book from 1963, Robinson gave an interpretation of Cauchy's writings. His result was: Cauchy's theorem is correct if we add one of the two additional assumptions: (a) the series is uniformly convergent or (b) the family $(s_n(x))_n$ of partial sums is equicontinuous in the interval.

Robinson's idea of translating historical mathematical texts into the language of Nonstandard-analysis impressed the philosopher of science Imre Lakatos (1922–74). In a lecture, which appeared in print only after his death, he claimed that "Cauchy's theorem was true and his proof as correct as an informal proof can be". According to Lakatos, the Cauchy Sum Theorem does therefore *not* need any additional assumptions to gain validity. Thus Lakatos contradicted Robinson.—The further development (i.e. that Lakatos was right in *quite another* sense from what he thought) was traced on pp. 130f.

A Satisfying Finish

It was the aim of my dissertation of 1981 to substantiate Lakatos' thesis. In contrast to him, I based my arguments on the Darmstadt version of Nonstandard-analysis. Today I am aware that my approach (see p. 132) was mistaken.

My doctor father Laugwitz was intrigued by my work, and it inspired him to undertake his own detailed studies of former analysts, *esp.* of Cauchy's works, thereby producing quite a few articles on the topic.

My studies of Bolzano's mathematics, which started in 1986, showed me the inadequacy of Lakatos' methodology in regard to the understanding of historical mathematical writings. As a consequence, I had to change my approach. In winter 1990, I turned again to the study of Cauchy's analysis, this time from a new perspective. Prof Dr Laugwitz tried to hinder my work with a maximum of vigour. After he failed, I lost his favour and consequently all local academic support (and sadly, more than that). The Darmstadt University showed itself incapable of an unbiased clarification of this issue until today.

If Prof Dr Laugwitz had been factually successful at his time (not only institutionally), this book could not have been written.

Literature

Aigner, M. & Ziegler, G. M. (1998). *Proofs from the book*. Berlin: Springer.
Lakatos, I. (1980). Philosophical papers (2 vols). In *Mathematics, science and epistemology* (Vol. 1). Cambridge: Cambridge University Press 1980.
Laugwitz, D. (1986). *Zahlen und Kontinuum. Eine Einführung in die Infinitesimalmathematik*. Mannheim: Bibliographisches Institut.
Luxemburg, W. A. J. (1962). Non-Standard Analysis. *In Lectures on A. Robinson's Theory of Infinitesimals and Infinitely Large Numbers. Lecture notes*. Pasadena, California: Mathematics Department, California Institute of Technology.
Luxemburg, W. A. J., & Körner, S. (Eds.) (1979). *Selected papers of Abraham Robinson* (Vol. 2). Amsterdam, New York, Oxford: North Holland Publishing Company.
Nelson, E. (1977). Internal set theory: A new approach to nonstandard analysis. *Bulletin of the American Mathematical Society, 83*(6), 1165–1198.
Palmgren, E. (1995). A constructive approach to nonstandard analysis. *Annals of Pure and Applied Logic, 73*, 297–325.
Palmgren, E. (1996). Constructive nonstandard analysis. *Cahiers du Centre de logique, 9*, 69–97.
Palmgren, E. (1997). A sheaf-theoretic approach for nonstandard analysis. *Annals of Pure and Applied Logic, 85*, 69–86.
Palmgren, E. (1998). Developments in constructive nonstandard analysis. *The Bulletin of Symbolic Logic, 4*(3), 233–272.

Palmgren, E. (2001). Unifying constructive and nonstandard analysis. In P. Schuster, U. Berger, & H. Osswald (Eds.), *Reuniting the antipodes—constructive and nonstandard views of the continuum 2001* (pp. 167–183).

Richman, F. (1998). Generalized real numbers in constructive mathematics. *Indagationes Mathematicae, N. S., 9*(4), 595–606.

Riemann, B. (1854). *Ueber die Darstellbarkeit einer Function durch eine trigonometrische Reihe.* cited from Weber and Dedekind 1953, pp. 227–271.

Robinson, A. (1961). Non-Standard analysis. *Indagationes Mathematicae, 23*, 432–440. cited from Luxemburg and Körner 1979, pp. 3–11.

Robinson, A. (1963). *Introduction to model theory and to the metamathematics of algebra.* Amsterdam: North-Holland Publishing Company.

Russell, B. (1919). *Introduction to mathematical philosophy.* In Georg Allen & Unwin, London (New York: The MacMillan Co.). www.gutenberg.org/ebooks/41654.

Schmieden, C. (1948–53). Vom Unendlichen und der Null. Versuch einer Neubegründung der Analysis. manuscript, unpublished.

Schmieden, C. & Laugwitz, D. (1958). Eine Erweiterung der Infinitesimalrechnung. *Mathematische Zeitschrift, 69*, 1–39.

Schuster, P., Berger, U., & Osswald, H. (2001). *Reuniting the antipodes: Constructive and nonstandard views of the continuum.* Dordrecht: Kluwer.

Schuster, P. M. (2000). A constructive look at generalized Cauchy reals. *Mathematical Logic Quarterly, 46*, 125–134.

Spalt, D. D. (1981). *Vom Mythos der Mathematischen Vernunft.* Darmstadt: Wissenschaftliche Buchgesellschaft. [2]1987.

Spalt, D. D. (1966). *Die Vernunft im Cauchy-Mythos.* Thun und Frankfurt am Main: Harri Deutsch.

Spalt, D. D. (2022). *Die Grundlegung der Analysis durch Karl Weierstraß—eine bislang unbekannte Konstruktion der natürlichen und der reellen Zahlen.* Berlin: Springer.

Strauß, E. (1880/81). *Weierstrass, Einleitung in die Theorie der Analytischen Functionen.* Universität Frankfurt am Main: Archiv. Sign 2.11.01; 170348.

Wallace, D. F. (2003). *Everything and more: A compact history of infinity.* New York, London: W. W. Norton (2010)

Weber, H. & Dedekind, R. (Eds.) (1953). *Bernhard Riemann, Gesammelte mathematische Werke.* reprint of the second edition 1892. New York: Dover Publications.

Author Index

© The Author(s), under exclusive license to Springer Nature Switzerland AG 2022
D. D. Spalt, *A Brief History of Analysis*,
https://doi.org/10.1007/978-3-031-00650-0

Subject Index

Printed in the United States
by Baker & Taylor Publisher Services